STUCK
MONKEY

STUCK MONKEY

THE DEADLY PLANETARY COST
OF THE THINGS WE LOVE

James
Hamilton-Paterson

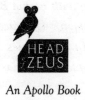

An Apollo Book

First published in the UK in 2023 by Head of Zeus Ltd,
part of Bloomsbury Publishing Plc

9 7 5 3 1 2 4 6 8

A catalogue record for this book is available from the British Library.

ISBN (HB): 9781803285528
ISBN (XTPB): 9781804545768
ISBN (E): 9781803285504

Printed and bound in Great Britain by
CPI Group (UK) Ltd, Croydon CR0 4YY

Head of Zeus Ltd
First Floor East
5–8 Hardwick Street
London EC1R 4RG

www.headofzeus.com

'Modern man no longer knows how to foresee or to forestall. He will end by destroying the earth from which he and other living creatures draw their food. Poor bees, poor birds, poor man.'

Albert Schweitzer, in a letter to a
French beekeeper, quoted by Rachel Carson

Contents

STUCK MONKEY

How the Monkey Got Stuck

I began writing this book in early November 2021 just as the COP26 climate summit in Glasgow was taking place. Despite the marked absence of major players like Russia and China, COP26's importance was hailed by most speakers and delegates as representing absolutely the last chance for 'world leaders' to make political decisions that might enable the planet to remain habitable by humans for the foreseeable future. And this despite everybody knowing that, in self-styled democracies at least, heads of state have very little power to dictate radical shifts in policy in a world where almost everything, and certainly energy pricing, is largely beyond their control because they have ceded power to transnational corporations. So much for democracy. Nevertheless, it was ritually agreed that the universal enemy is global warming and all remedy appears to hang on reducing reliance on fossil carbon sources, chiefly natural gas, coal and oil: all still vital enablers not just of the global economy but also of life itself for billions of people. On 1 November the BBC's website headlined just how serious it thought the matter: 'World temperatures

1

are rising because of human activity, and climate change now threatens every aspect of human life. Left unchecked, humans and nature will experience catastrophic warming, with worsening droughts, greater sea level rise and mass extinction of species.'

The grab-bag of the summit's journalistic clichés was familiar enough. An editorial without the word 'catastrophic' in it was rarer than hen's teeth. Many including the then-Prince Charles gave their considered opinion that we were now in the Last Chance Saloon. Furthermore, there were no genies to be plucked from magic bottles, no wands to be waved. Now everything depended on technology. This meant switching to alternative, 'clean' sources of energy such as wind, wave and solar power to generate electricity, together with newly efficient batteries for storing it. At the same time we would need to develop alternative, zero-emissions fuels such as liquified hydrogen for engine-driven transport. The usual ghostly silence surrounded nuclear energy, the cleanest of them all, with zero carbon emissions when generating electricity.

It is now almost immaterial whether human-induced global warming is actually taking place. For years its reality has been relentlessly endorsed by the great names of the Earth sciences and most of the world's press, increasingly coupled with the entirely unscientific word 'consensus'. Today, anyone can see its apparent effects everywhere. Global warming has the self-fulfilling properties of UFOs, such that almost everybody now believes 'there must be something in it'. To make quite sure we all know what to think about the way the world's great and good attended the climate summit in Glasgow, newspapers obligingly pointed the moral with headlines like 'Hypocrite Airways? Jeff Bezos's $75m Gulf Stream leads parade of 400 private jets into COP26 including Prince Albert of Monaco, scores of royals and dozens of "Green" CEOs'.[1] The image of those 400 private aircraft crammed into the parking lot of the Last Chance Saloon provided a welcome touch of satire. Nor had it gone unnoticed that US president Joe Biden arrived from a G20 Summit in Rome where he and his entourage drove around in an eighty-five-vehicle convoy. The absurd self-importance of these so-called 'world leaders' consorts oddly with the global damage caused by what they represent. Bezos's Amazon bestrides the world far more like an octopus than a colossus; while Commander-in-Chief Biden's US military with its 4,800 bases in 160 countries has been named as 'the world's single largest consumer of oil and [...] one of the world's top greenhouse gas emitters'.[2]

So it seems we no longer need the significance pointed out to us of how exhaust fumes are playing their part in tornadoes, bushfires, droughts or horrendous floods such as those in Germany's Ahr valley in July 2021, even though the evidence suggests that those particular floods were to some extent the direct result of culpably bad local planning that permitted extensive building on what should have remained riverine wetland. At that point the Ahr is about to flow into the Rhine and its valley is historically known to have flooded causing fatalities, such as in 1804. But planners have short memories and instinctively feel that modern know-how can anyway deal with threats that were once a danger to preindustrial societies. The overall message of COP26, with its besuited 'world leaders' and impassioned teenaged saints delivering sermons, was that the gloomier forecasts may yet be averted by technology. The purpose of this book is to show this to be little more than brainless positivity. Naïve optimism is common to many nations but in its pathological Pollyanna form is the besetting sin of the United States, with that country's conviction that technology will always come to humanity's rescue with undreamt scientific breakthroughs. (This despite cumulative scientific breakthroughs down the centuries being at the root of the planet's present plight.) The main problem with this sort of wishful thinking is that it conveniently mistakes symptoms for the disease itself. Global warming itself is merely a symptom; it is not the malady. We need to make a radically different assessment of what lies at the root of global warming. The planet's real disease is us – or, for present purposes, The Monkey.

There is a story – whether fable or true – that some people hunting monkeys in the jungle once devised a simple trap that proved remarkably effective. It was nothing but a stout glass jar with a comparatively narrow neck, such as might once have contained a local delicacy such as salted plums. Into this they put a large juicy banana and then went back through the jungle to their village for lunch. On their return they discovered that they had indeed caught a monkey. Having plunged its paw into the jar to grab the banana, the creature found its fist was now too bulky to fit through the jar's neck. Unfortunately for the monkey, its yearning for the banana was too overwhelming to make it let go of the fruit, withdraw its paw and then simply upend the jar. On seeing the hunters arrive, even panic failed to lend it the wit to smash the jar against a boulder or a tree trunk and make off with the coveted banana. Instead, its intelligence had simply fused and all it could do was jump up and down in a frenzy with the heavy pot on its paw. The hunters easily took the wretched animal to put it into a still larger pot.

The Monkey is of course us, and the way we are paralysed by our inability to relinquish or even change our way of life and its consumer goodies on which the global economy depends. True, in the spirit of the times citizens of developed countries who care enough and can afford it buy hybrid cars, instal solar panels and (electrically powered) heat pumps, and shun plastic bags. This seems like mere virtue signalling if these same people still buy ever more things on Amazon Prime, shop for akebi fruit from Japan (when available in Waitrose), and book a trek to Machu Picchu or a diving holiday in the Seychelles.

In time our machines may well be tamed and some of their pollution abolished. With luck even a patch of The Monkey's jungle can be saved. Yet periodic jamborees like COP26 and the desperate quest for carbon neutrality are expensive farces. By far the greatest threat to the planet's health comes from the inexorable demands made on its limited resources by its human freight and by our obsession with burning things. The environmental historian Stephen J. Pyne has coined a word to describe our era: the Pyrocene. 'A global regime of burning: coal and oil, agricultural land and forest, bush and wetland, most of it planned... Wherever you look, the earth is in flames. The residue is carbon monoxide, nitrogen dioxide, ozone, black carbon, sulphur dioxide, and the particularly toxic grouping of small particulate matter known as PM2.5. Everything we burn, we breathe.'[3]

Countless poor souls have experienced the awful shock of hearing a doctor confirm they have contracted a virulent form of cancer. The blow may be softened by a cheerful prognosis or ritual words of hope; but one morning in a well-lit hospital office with a potted Swiss cheese plant on the windowsill the sufferer is brought face to face with the unavoidable reality of their own death. No longer might it lie in wait a hopeful forty years down the line, by which time a cure surely will have been found. No longer is it something that only affects other people. Suddenly, it really is now your turn.

The immediacy of this bowel-draining shock is exactly what we all ought to be experiencing over our own environment's cancer diagnosis. The signs have been clear for decades but political flummery, outright denial and American-style positivism have always kicked in to enable us to carry on much as usual. That and the iron laws of the global market. The Covid pandemic brought about more actual change to people's lives worldwide than any measure ever taken to protect the planet's health. Ever since, the political priority everywhere has not been to rethink and start afresh more sensibly, but to 'get back to normal' as soon as possible.

The overwhelming difficulty is that there is no unity. Economists and politicians tend to fret only about their own country's – or bloc's – performance, or about anybody else's only insofar as it affects their own. Had we had the sense decades ago to realise that we should be worrying primarily about the planetary economy, we might not now find the global environment facing a cancer prognosis gloomy enough to qualify as a death threat. Yet the reality is that we are as far now from instituting any practicable world governance as we have ever been. No measure that makes significant inroads into our rabid global consumerism stands the faintest chance of being universally agreed, let alone implemented and policed. We are still parochial in our outlook, still too devoted to the national lifestyle we already have or the one we aspire to. Our attempts at global change are constantly undercut by disparate influences as the biggest countries, industries and corporations vie with each other like dinosaurs bellowing in a swamp for superpower dominance while the lesser take sides or demur. Not only that, but the decriers of anything that casts doubt on the inherent rightness of the marketplace always produces the ritual, even belligerent, prognosis: Sure, things may look a bit grim now but have faith: the market knows best. Gordon Gekko lives! Greed is good! You'll see. Things will straighten themselves out just as they always have...

Only this time they won't. It is partly the nature of the marketplace itself that has led to our world's terminal illness; but it is also the way we treat that marketplace as having a godlike inherent rightness. 'While fetishes made by African priests were denigrated as irrational, the fetish of the capitalist marketplace has long been viewed as the epitome of rationalism,' as Anna Della Subin wrote recently. 'Abstract European social theories are no more universal, eternal "truths" than African gods are.'[4] What has thrown everything permanently out of kilter is the pursuit of economic boom through unlimited (if geographically unbalanced) consumerism, rather than an aspiration to a modest standard of living that might have made possible a reasonable life for all of Earth's nearly 8 billion people.

This is not to preach. The truth is, we can't help it. We are The Monkey. Each one of us is hardwired by evolution to strip trees bare of fruit when we find them: to survive by grabbing as much as possible of everything in case of later famine. We will only ever act to the benefit of our own immediate tribe, and the economic system we have fashioned merely reflects this. It is the legacy of countless millennia of hunter-gathering, of hand-to-mouth survival that today leads to panics when people clear supermarket shelves of the last toilet roll on the basis of a mere rumour of shortage. There was a story of a man queuing for hours at a petrol station during a recent tanker drivers' strike, getting his petrol and then driving nine miles to collect his daughter from school and deliver her home. After that he promptly drove back to the petrol station and rejoined the queue to top up, having already wasted upwards of three hours of his life in a queue. *Get it while it's still there.* To call such stories examples of FOMO (fear of missing out) or greed is simply to moralise a genetic imperative. We who are briefly alive at this moment are the glutted survivors of our species and stand on an alp of corpses well over 200,000 years high. It is best not to enquire too closely into who of our forebears were the best grabbers, nor who among us now are dying of hunger. There will soon be even more people standing on us.

And that is the second direct cause of the planet's sickness: precisely those 8 billion hungry mouths and grabbing hands. In 1950 when I was eight years old the world population stood at 2.5 billion, and if the planet as a whole was in any danger it was considered far more threatened by thermonuclear war than by population pressure. Today, the same planet is host to over three times that number of people while the likelihood of nuclear destruction still exists, although maybe with a little less urgency. The more immediate threat to planetary survival is that figure of 8 billion people almost all of whom, thanks to modern technology, are perfectly aware of the sort of things on offer in more privileged societies and understandably want them for themselves.

That is not going to happen. Not only are the planet's natural resources not infinite, they are already seriously compromised by overexploitation and climatic change. No matter what we do now, populations are already on the move and countries will become extinct. This will either be because of rising sea levels (the Maldives); rising temperatures (Mauritania, the Gulf States); or because a long succession of disgustingly corrupt, inept and brutal governments at home drive their youngest and best potential wage-earners to emigrate. As I was writing this a great horde of people from South America, particularly Venezuelans, was stumbling its way on foot through the Panama jungle towards the Promised Land of the United States. En route through the isthmus they would gather further pilgrims from other failed states (Nicaragua, Honduras, Guatemala). Several thousand miles to cover, often with small children, at the mercy of venomous snakes and equally venomous border officials, thieves and brigands. At the same time other desperate people are trekking northward from sub-Saharan Africa towards the Promised Land of Europe. The bones of many of them litter the desert or drift down to the Mediterranean sea bed while the survivors overwhelm islands and refugee camps where their progress is officially blocked. Neither in Europe nor the US is there a clear agreed policy on how to deal with these wretched people, any more than there is in the long-established refugee camps scattered in Jordan and throughout a Middle East devastated by wars we largely initiated. *Not our tribe.* As of writing, Britain was attempting to export migrants to Rwanda in a chartered aircraft. There exists no political body with the overall clout to force any of the Promised Lands to take these people, nor anyone that wants them. The sole body that might have agreed a unified policy, the European Union, has ignominiously failed and has allowed ruthless gangs of people smugglers to pack people into rubber boats for the often-lethal Channel crossing. Elsewhere, millions of refugees will remain and fester in their improvised tent cities, growing ever more disaffected and radicalised. The situation can only get worse.

It is true that for some years now there has been growing awareness that our consumerist way of life might be having major climatic consequences. Better-heeled and more thoughtful people do now sometimes think twice before taking long flights, aware of such things as carbon footprints and vapour trails. In Sweden in 2018 a movement known as flygskam or flight-shaming began, although since flying anywhere was soon made far less possible by the Covid pandemic it is not clear how effective this campaign may yet be. Devout – often Republican – capitalists tend to mock such middle-class consumer renunciation as borderline socialist. They argue that long before The Monkey evolved, the planet's archaeo-history shows clear evidence of abrupt climate changes of unknown origin: sudden ice ages and rising and falling sea levels. They reject an overwhelming majority of Earth scientists who claim that current global warming is directly related to (among other things) the pollution of the atmosphere with carbon dioxide and other chemicals that has steadily increased since the Industrial Revolution in the late eighteenth century. Science does not agree with such devout capitalists any more than I do. Even the smallest intelligence must acknowledge that in a closed system such as our planet's, surrounded as it is by the vacuum of space, the massive and ever-increasing consumption of fossil fuels over the last two centuries – the gaseous by-products of which wind up in the air and the oceans – can't not have had an effect on the climate as well as on the rest of the natural world. Where else could these pollutants have gone? Their history is now legible in black-on-white as soot written deep into snowcaps and glaciers.

Once this commonsensical notion achieved global currency people have tended to think simplistically but understandably in terms of factory chimneys and exhaust pipes. Smoke is a visible sign of pollution and it is logical to concentrate on such sources. Thus cars and power stations have been relatively easy to demonise while science and technology devise magic wands to wave over them (e-vehicles, solar and wind power) to make them environmentally friendlier.

'Environmentally friendly' is just one of many delusional feelgood phrases beloved by ad agencies behind which concerned and decent citizens can take shelter, even if in the final analysis the environment would frankly be far better off without any of us, friendly or otherwise. Still, even if we humans can no longer quite believe in our own inherent cuddliness, we at least consider we have the intelligence to care about our future. Save The Planet! No crusade ever marched under a nobler banner, nor waved a more fatuous one. Our environment may be cancerous but the planet itself is not remotely endangered. It has already survived for approximately 4.5 billion years, its face radically remodelled from time to time by asteroid impacts, massive volcanic eruptions and continental drift. In due course it will cheerfully shrug off its latest dominant species and carry on producing new lifeforms until it doesn't. What today's crusaders mean is entirely self-interested: Save the planet in its present form that so admirably suits me-me-me. Save also the various plants and animals we see (even if few people can put a name to them). Most of us will never see an aardvark or a baobab tree or a giant squid and probably don't feel like making the effort to go and look at one since pictures and information are readily available via the Internet. But we like to know they're still there, along with Siberian tigers and orangutangs and all the rest. News of species going extinct seems always to carry an ominous, doom-laden finality even though it has been estimated that at least 99 per cent of all the species that have ever existed on Earth are long extinct. Extinction is just the flipside of Evolution: both go hand in hand. Long after we're gone the Earth will carry on producing astonishing new species for the next billion years, or until such time as our sun burns the last of its hydrogen and gradually turns into a red dwarf, expanding until it engulfs much of the solar system. At least that's one thing none of us need ever worry about.

Our present worry ought to be that The Monkey's way of life is radically incompatible with the health of the planet. Unfortunately, it is human beings' overwhelming urge for gratification and pleasure that makes any real change in our habits so unlikely. After a century or so of being sold the American model of consumerism, few people are much inclined towards self-denial. As the long-drawn-out Covid pandemic showed, people need – or feel they need – compensation or reward for upset, for the anguish of personal loss, for a disrupted lifestyle that obliged them to try and earn a living from a too-small home shared with young children who needed endless pacifying as well as home schooling. Comfort eating, comfort drinking, comfort drugs, comfort buying, comfort violence. No matter how grim the circumstances of foreign travel were in the late summer of 2021, thousands still thought it worth the effort to snatch ten days or a fortnight on a Mediterranean beach, and to hell with the environment. Meanwhile the airlines and cruise companies were, and still are, practically begging people to return to their pre-pandemic travel habits and so are hotels, resorts, pubs and clubs. The entire global economy is firmly rooted in restless desire.

It is quite obvious that the majority of people will never voluntarily forgo their jollies unless they can clearly see it will be the same for everyone else, regardless of nation, class, wealth or political clout. This is human nature; and recognising it was largely what made rationing the success it was in Britain in the Second World War. A handful of hoarders and spivs apart, it was widely accepted as being no fun but necessary and fair. But what one country can reluctantly agree to under the imperatives of wartime cannot be made to work across the world's 195 utterly disparate countries. Quite apart from the problem of Country A undergoing cutbacks when it can see that Country B is flagrantly indulging itself, the times are now quite different. The immediate threat of war, backed up by foreign warplanes droning over your house, has about it a personal urgency that a mere environmental threat to the planet entirely lacks.

The purpose of this book is to look at certain ingredients of our treasured consumerist way of life that we like to imagine don't really have consequences for the planet after all. What could be wrong with 'just wanting to live a normal life?' What more innocent than gardening, or wanting to keep a pet? What more essential than having a mobile phone, or more unassailably righteous than playing a sport or going to a gym? We naturally turn a blind eye to what we would rather not know. In our developed world we virtually none of us make or grow the things we need. We almost all lack the time, the space and the skills. We now buy practically everything ready-made and increasingly so online. It all has to be manufactured from raw materials that leave behind them sundry trails of pollution in one form or another, as does delivering the goods to our doorstep and disposing of them when we tire of them and dump them. As we tip them into a skip the more reflective among us may try to think back to wonder if at the time we bought them we really needed or even wanted them. We probably can't remember. There have been so many things since... But then, the essence of the consumerism that underwrites the world's economy is to blur any distinction between 'want' and 'need'.

From the planet's point of view, and in order to recover from its current sickness, it both wants and needs fewer and better-behaved humans. On present showing it's not going to get them. The prospect of lessening our environmental impact does not automatically diminish with each new technological breakthrough. Rather, it all amounts to an elaborate game of pass-the-parcel. Possible remedies for the steadily growing consequences of our way of life for a finite planet are simply handed on from one technology to the next, each of which can for a while appear plausibly cleaner than its predecessor until its true hidden cost is uncovered. As every physicist knows, where energy is concerned there is no such thing as a free lunch, and nor can there ever be.

Eat Your Pets!

The function of pets is to be humanity's smiley emoticon. We have already trashed a major part of the wild, but by pampering the tame we feel we are somehow doing our bit to restore the balance.

In reality we are simply adding more damage. This is largely because in the space of a few decades since the Second World War the keeping of pets, like so many other formerly innocuous human activities, has fostered major global industries. If one were to portray Big Pet Food's demand for meat alone as a separate country it would now rank as the world's fifth-largest consumer behind only Russia, Brazil, the US and China.[1] One authority estimates that pets now consume around a fifth of all the world's meat and fish.[2] The production of red meat accounts for 6 per cent of all greenhouse gas emissions. Global beef production alone produces emissions roughly equal to those of India and requires twenty times more land per edible gram of protein than do protein-rich crops such as beans.[3]

Yet the true environmental cost of the global pet industry goes far beyond that, even as it has become a topic whose emotionality illustrates just how unlikely is any prospect of meaningful change. Particularly in Britain, events such as the killing of Geronimo the alpaca in late August 2021 and Pen Farthing's simultaneous efforts to bring 200 ex-military dogs to Britain in preference to people on the precious last flights out of Kabul reveal a potential for simplistic public emoting over issues of animal welfare that is hopefully unmatched anywhere else. Tom Tugendhat, the Tory MP and ex-serviceman who had been struggling to get his old army interpreter out of Afghanistan and into the UK, commented trenchantly: 'We've just used a lot of troops to get in 200 dogs; meanwhile my interpreter's family is likely to be killed. When one interpreter asked me a few days ago, "Why is my five-year-old worth less than a dog?" I didn't have an answer.'

In today's Britain a single animal's cause, once taken up by the popular tabloids and the creature given a name, can quickly escalate into a political issue. In the summer of 2021, after two years of legal cases and injunctions, more than 140,000 people including then-prime minister Boris Johnson's own father, Stanley, signed a petition for Geronimo's stay of execution. Despite a group calling themselves The Alpaca Angels being hastily formed with the avowed intent of thwarting the animal's 'murderers' by using decoy alpacas, Geronimo was duly dispatched in line with the policy of the Department for Environment, Food and Rural Affairs (DEFRA) to try to eradicate bovine tuberculosis in the UK. The animal had twice tested positive for the disease using what the Animal and Plant Health

Agency (APHA) called 'highly specific tests'. Four months later APHA admitted 'it was not possible to culture bacteria from (Geronimo's) tissue samples taken at postmortem examination, meaning that it will not be possible to carry out whole genome sequencing in order to try to understand how the alpaca caught the disease. This does not mean the animal was free of bTB infection'.[4] No; but it will probably guarantee further shrieks about its supposedly untimely end.

Britain's broadsheets can on occasion be slightly more adult and toughminded than its tabloids and among *Guardian* readers there may well be a growing awareness of the vast annual toll of birds, rodents and other wildlife taken by Britain's cat population, both domestic and feral. Yet despite an increasing number of people (very often children) who are mauled and even savaged to death each year by pet dogs, the tabloid world remains overwhelmingly dominated by a general celebration of furry cuteness that can do no wrong, and woe betide anyone with the temerity to claim otherwise. In short, and as always, the feelgood factor overrides intelligence and it becomes ever more difficult to think straight about animals in general, let alone pets. The majority of pet owners are surely unaware of the far-reaching environmental consequences of their companion animals, whether out of active denial or from a warm, blurry feeling that having animals about the home somehow brings the household closer to Nature and hence exonerates it from having to acknowledge an environmental cost. It may be that the pleasure, companionship and solace pets can undoubtedly afford significantly outweighs their disadvantages. In line with this they ought perhaps to be reckoned economically and socially as differently shaped, fur-covered children. As such, they achieved star billing during the Covid lockdown.

RETAIL PET THERAPY

It was retail therapy that enabled millions of citizens to survive the lockdown: itself a revealing pointer to one of the mainsprings of consumerism. For many people it is a near-instinct to 'reward' themselves for upset or unhappiness by buying things, and by no means only comfort food. Thousands in lockdown went online and ordered luxury items like pizza ovens. One in twenty adults bought a hot tub, with 36 per cent of them apparently now regretting it. 'During the pandemic we spent on average over £1,200 each at a collective cost of £57.6 billion. But £6.6 billions' worth of those purchases are no longer used. The most bitterly regretted impulse buys include home gym equipment, gaming gadgets and musical instruments.'[5] Notably absent from this list are pets, which are to a large extent treated as if they were live stuffed toys. For, presumably to break the domestic tedium and distract the kids, there was also a sudden large increase in pet ownership, much of which would soon be repented.

Figures published in March 2021 by the Pet Food Manufacturers' Association showed that 3.2 million UK households had acquired a pet since the start of the pandemic, many of them dogs that sent 7,386 people to hospitals in England with bite injuries.[6] As a result of the sudden demand, the average price of puppies doubled to a scarcely believable £1,900 and dognappers and black marketeers inevitably plied their ruthless trade. And this for an animal that, unlike a cat, hasn't the brains even to come in out of the rain but will just stand there, whining and dripping. Much of the supply came from illicit breeders, predominantly in Ireland, who exported the puppies via Belfast to Scotland, Wales and England in a business that one animal welfare officer described as 'the new drugs trade'.[7] Predictably, once lockdown had eased many of these newly acquired pets were simply abandoned. Some owners brought them to animal shelters claiming they were strays they had found, but in many cases shelter staff were not fooled since most genuine streetwise strays can be quite hard to catch. By early 2022 animal shelters were saying that most of these 'fake strays' now had medical problems. According to the Dogs Trust's figures over 47,500 dogs were abandoned by their owners in the UK in 2020 alone. Those not run over mostly ended up in council pounds and more than 5,000 were later put down.[8] The daily executions of Britain's formerly adorable but suddenly unwanted pets have proceeded ever since. Because the animals are unnamed and their cause is not endorsed by the tabloids nobody much cares. A related problem has been the sudden upsurge in cases of dog bites resulting from clueless owners without the first idea of how properly to train and socialise their animals.

A similar mass-abandoning of pets was seen after the financial crisis in 2008. Such things are predictable because many potential owners fail to properly cost the long-term implications of pet ownership and pets are expensive. The current UK cost of keeping a dog in its first year is generally estimated at £1,800. A recent study found that the average cost of a 'designer' dog is £2,000 in the first year alone. Excluding the animal's initial price, this figure is made up of insurance (Third Party as well as for the animal itself), food, and various veterinary fees that would include spaying and microchipping. The website of Britain's PDSA estimates a dog could cost £30,000 over its lifetime. This figure does not include vet fees but does include insurance.[9] A simple vet's check-up for even a small dog can cost £80–£100: a procedure that many insurance companies won't even cover.

In the UK over a quarter of all households own one or more dogs and 18 per cent have cats. This translates into large sums of money for the pet industry, which of course involves far more than just food. By 2020 the global pet care market surpassed $232 billion and was poised to expand: one of the few industries to have reliably grown during the Covid pandemic.[10] Out of that total in the same year the US easily headed the world pet food market at $31.144 billion. The UK came second with $5.262 billion.[11] On top of that the vaccinations, ID 'chipping' and pharmaceuticals such as flea powder and antibiotics must usually be added the cost of kennels, grooming, walking and insurance. This explains how the money spent on pets in the UK each year (food, grooming, accessories and veterinary) was calculated in 2019 to be £6.5 billion and will have risen significantly since then. Pets are very big business.

BIG PET FOOD

Where the global pet food industry and its environmental impact is concerned, sources like Public Library of Science's peer-reviewed, open-access journal *PLOS One* have recently published some useful research, even if much of it is specific to the United States. Nevertheless, today's Big Pet Food industry is thoroughly international and most findings will apply equally to wealthier European, South American and Asian countries, if not worldwide. The United States believes it has approximately 163 million non-feral dogs and cats but this is probably a gross underestimate. Having analysed the amount of meat in the most popular pet foods, the geographer Gregory Okin of UCLA's Institute of the Environment and Sustainability calculated that America's dogs and cats are consuming 19 per cent of the calories eaten by the country's 321 million people. Gone are the days when pets were largely fed on their owners' leftovers bulked out with bread crusts or whatever came to hand.

After the Second World War, in the UK many foods, and especially meat, were still rationed. With the ending of enemy submarine attacks, British whalers went back to work and for a while we ate whale meat, often as a stew: a dark, oily substance. In time more imported beef and lamb reached us from Australia and New Zealand and whale meat went to feed the dog, plus the scraps and bones that all high-street butchers had always put by. Fish followed a similar pattern. At first we ate tinned snoek, a widely disliked South African fish, and then progressed to coley, the least expensive of the cod family until that, too, became cat food. I don't remember buying tinned dog or cat food for our thriving animals until the very late 1950s when we did switch to tinned pet food, more for our own convenience rather than out of concern for our pets' diets.

Today's pet food scene could scarcely be more different and what is on offer amounts to a gourmet's banquet that during the war any human family would have gobbled up with gratitude. Pound for pound, today's pet food contains more meat even than the average human diet does and Okin was able to recalculate and find that between them, modern pets eat a quarter of all the meat the US consumes. He further determined that merely producing this much meat each year entails the generation of 64 million tons* of carbon dioxide: equivalent to 13.6 million car exhausts (the US Environmental Protection Agency's figure).

In 2018 it was estimated that the global pet food industry is responsible for a quarter of the total environmental impact of all the world's meat production in terms of land use and the consumption of water, fossil fuels (coal and oil), phosphate fertilisers and pesticides. How many pet owners today equate

---

* Both 'ton' and 'tonne' have been used throughout the book. This is down to the references quoted, 'ton' usually appearing in British and US sources, and 'tonne' in those that use the metric system. Nor are the Imperial and US tons identical.

the meat in the chunks they spoon out of a tin with the clearance of Brazilian rainforest in order to grow soya as cattle food? Back in 2009 researchers estimated that a medium-size dog has a similar carbon footprint to that of a large SUV. This figure will certainly have gone up since then to take account of today's pet foods containing an ever-increasing proportion of meat. Perhaps the main reason for this is the widespread belief that since all dogs are presumed to be descended from wolves, they ought to eat like them. This notion is the canine equivalent of human faddists' 'paleo' diet, and is understood to mean a very high proportion of meat. This is what www.veterinary-practice.com has to say on the subject:

Overall, diets that are marketed as 'natural', 'evolution-arily correct' or 'biologically appropriate' assume that we know the diet domestic dogs and cats have evolved eating and presume that feeding such diets will promote longer and healthier lives. These are very attractive arguments, but a diet being more natural or closer to the ancestor's does not imply optimum nutrition, since animals in the wild eat what is available and the lifespan of wild canids and felids is shorter than that of their domestic counterparts. The goals of nature/evolution are also different from pet owners' goals, and feral animals can die young due to broken teeth, malnutrition and parasites.

Certainly wolves in the wild, depending on the season, become opportunistic omnivores and happily extend their diet to include fish (typically salmon), berries and fallen fruit.

In order to supply pet dogs with a largely meat diet the industry increasingly looks to unconventional sources besides the mechanically separated meat (MSM), mechanically separated poultry (MSP) and the 'pink slime' that already goes into human processed food. It includes grinding up udders and lungs as well as meat from alligators, ostriches, Asian carp and insects, afterwards bulking it out with rice bran, vegetables, pork protein powder and 'spray-dried animal plasma' (SDAP). This latter is a powder made from the blood of animals slaughtered for human consumption. Rich in various proteins, it is much used in feed for poultry and pigs as well as for pet food, where it is particularly useful as a binder for the 'meaty chunks' touted on tins and pouches of dog and cat food. The processing of these ingredients is energy-heavy. At a pinch (and their owners' beliefs permitting) dogs can live quite well as vegetarians, whereas cats cannot if they are to remain healthy. This is because feline metabolism actually requires a certain percentage of meat – preferably raw – which presumably accounts for cats' undiminished instinct for killing vast numbers of birds and small mammals worldwide, as well as species already threatened by other things (moths, butterflies, lizards). UK cats alone are estimated to kill up to 300 million prey animals a year, mostly songbirds.[12] They are also believed to kill a quarter of a million bats annually, which is why many naturalists beg cat owners to keep their pets in at dusk when bats are at their most vulnerable. This instinct for indiscriminate murder is a black mark against felines to match the gross ecological impact of dogs on top of the annual tally of bites and fatal maulings they inflict on humans.

It is no coincidence that the top two biggest pet food companies in the world (based on their 2019 revenues) are Mars Petcare Inc. at $18.085 billion and Nestlé Purina PetCare with $13.955 billion. Both companies' fortunes were founded on confectionery and human foods and enormous sums are spent annually on research into what the industry calls the craveability of pet food. The aim is always to produce food that the animals – and especially picky cats – will find moreish. The discovery that cats favour palatants like umami and kokumi was something of a breakthrough for the industry. Dogs present a more difficult challenge since, as any owner watching their pet gleefully rolling in roadkill or vomit knows, canines have a predilection for such things. As Marion Nestle, professor emerita of nutrition at NYU, puts it: 'Carnivorous animals eat faeces. They like strong animal odours and pet food manufacturers have a really difficult time because they have to make it disgusting enough that the animal will eat it, but not so disgusting that the owners won't buy it.'[13] To this end complex chemicals like putrescine and cadaverine may be added to dog and cat food. As their names suggest, both are produced by the breakdown of proteins when a dead body rots. They have smells that raw meat-eaters like cats and dogs find irresistible, while most humans find them vile and vomit-inducing.

The research constantly conducted into food for every kind of pet shows how much highly competitive care and thought go into all aspects of the matter. Little is overlooked. Should your cat live in an Islamic household it can eat Muezza Pure Halal Formula cat food made from 'human-standard' ingredients in Lancashire. It has long been obvious that great numbers of the world's pet cats and dogs are far better fed and cared for than are billions of the planet's poorest or unluckiest humans. Whether or not this feels morally wrong or merely ironic is a matter for the individual; but it nevertheless has a considerable daily impact on the planet itself. Children's good behaviour is often bribed or rewarded with sweets or an ice cream and many pet owners likewise give their animals 'treats' and chews which themselves constitute a highly lucrative niche in the pet food market. In an Austrian garden centre in April 2021, shelved next to tins of dog food generically labelled 'Kangaroo' and 'Horse', the following packets of dog 'treats' were on sale: Crunchy Mackerel with Raspberries; Ostrich & Blackberries; Sardine & Wild Garlic; Venison with Strawberry Leaf; Crunchy Wild Boar with Rosehips; Quail with Oregano; Duck with Timothy Grass; Buffalo with Rose Blossom. (These are all products of a Czech company called Carnilove: producers of 'holistic feed' for dogs and cats that is exported to seventy countries.) One would hardly be surprised to find exactly these flavours for human 'nibbles', snacks or crisps in a UK supermarket like Waitrose. Any irony surrounding these gourmet dog delicacies in a Europe of food banks, refugee camps and undernourished children is merely one of the delicious ironies of late capitalism.

As I wrote this, hundreds of bedraggled Middle Eastern and Afghani refugees and their children were pinned in freezing temperatures against the soldier-patrolled barbed wire that prevented them getting from Belarus into Poland. They would surely have happily settled for a supper from the range of tinned dog foods made by the Bavarian company Terra Canis (Latin for 'It's a Dog's World', more or less) that prides itself on all its ingredients being '100% human grade'. They could have chosen between Ostrich with Parsnips; Buffalo with Millet; Tomato and Papaya; and the 2021 Christmas special: Winter Deer with Chestnut, Tangerine and Winter Spices. It all seems weirdly aimed at the human rather than the canine palate. Left to themselves, most dogs will cheerfully eat practically anything: roadkill, gizzards, wattles, sphincters and, of course, homework.

In the über-Green cause of Want Not, Waste Not it ought to be standard practice that the daily heaps of corpses of all those euthanised 'fake strays' are fed back into the hoppers of the pet food industry to become a canine version of Soylent Green. After all, it might ensure that a few square metres of Amazonian forest were left uncleared. It might also be time to resurrect the practice of the two World Wars when stray cats and dogs vanished not-very-mysteriously from the streets of Britain's cities at a time of stringent meat rationing. Dog is no more difficult to cook than lamb and is less smelly, although it does require practice to produce recipes of gourmet level. Cat cooks like rabbit but is drier.

COLLATERAL DAMAGE

In short, on grounds of diet alone the huge quantities of meat and vegetables – all requiring energy-intensive raising, processing, packaging and distribution – make the world's pet cats and dogs major daily contributors to climate change and environmental damage. To this can be added other steep environmental costs such as those incurred by their medication and the disposal of their excreta. Like SUVs, pets also have exhausts: in the US alone cats and dogs are currently estimated to produce an annual 5.1 million tons of faeces. This is almost a third of what the human population of America produces and each year generates about 64 million tons of methane and other greenhouse gases. In the UK dogs alone produce a thousand tons of excrement every day, much of it dutifully scooped into plastic bags, not all of those biodegradable. (In 2020, by a happy coincidence, this annual mass of 365,000 tons of dog poop also equalled the exact total tonnage of Nutella eaten worldwide.)

Let us address the savoury topic of cat poop first.

I love little Pussy,
Her coat is so warm;
And if I don't hurt her
She'll do me no harm...

You wish. Though less in terms of sheer tonnage than dog poop, cats' excreta are potentially harmful to owners and wild animals alike. This is because although many creatures, including dogs and humans, can be hosts of the parasite Toxoplasma gondii, it can only reproduce inside cats. T. gondii infection can cause the serious – and occasionally fatal – illness toxoplasmosis. These days T. gondii is one of the commonest parasites in developed countries, so much so that it has been estimated that up to half the world's human population carries it in the form of a latent infection. It might initially present in a patient as a bout of mild, flu-like symptoms before the parasite lies dormant in its host. It has been accused of producing subtle neurological and behavioural changes. This remains unproven; but research is ongoing to investigate this parasite's possible connection with both schizophrenia and bipolar disorder. T. gondii has been detected in nearly all wild bird species, and although some birds (including owls, hawks, sparrows and turkeys) prove resistant to it, others like pigeons are very susceptible and those it nearly always kills. As a recent report in *New Scientist* made clear, 'Wild animals living close to cities are more likely to carry the parasite that causes toxoplasmosis, and house cats may play a major role in spreading it to wildlife.'[14] While cat poop may potentially be no more harmful than dog excrement, cats that are kept strictly indoors are likely to be less widely contaminating – at least until it comes to disposing of their trays of kitty-litter. However, the majority of cats do roam outside and will of course readily urinate and defecate in neighbouring gardens, vegetable patches and planter boxes.

Cat-lovers often congratulate the animal on appearing to be more fastidious by making some effort to bury its leavings, unlike dogs that witlessly perform the barely token gesture of scratching randomly backwards before trotting nonchalantly off. Yet some of cats' T. gondii parasites will, via rain and river runoff, find their way to the sea where they can infect otters, causing sickness and death. Cat owners usually hope their pets will confine their excretory habits to trays of clay-based cat litter. Yet even this inert, absorbent substance causes environmental damage. In its original state the clay is generically known by its geological name bentonite and most of it formed millions of years ago from layers of volcanic ash on prehistoric sea beds. When it is conveniently accessible beneath the soil on dry land it is strip-mined (never a very eco-friendly way of obtaining raw materials). Bentonite is used in a number of industrial processes and a quarter of a yearly total of 25 million tons of it is mined in China. It is widely used to make cat litter, being odourless and highly absorbent. In the US alone an annual 2 million tons of contaminated cat litter winds up in landfill. Because clay is not biodegradable (having eons ago reached its final, geological state of decomposition) it will remain there unchanged until Doomsday, long after the cat poop itself has decomposed and leached away. The disadvantage of clay-based cat litters is that they require extensive processing, packaging and transportation. 'Green' they aren't.

As for dogs' excrement, this – like cats' – may well contain the parasite T. gondii but more attention is generally paid to the dog roundworm, Toxocara canis. This is very common and is passed on via the worms' eggs in dog faeces. This transmission is frequently accomplished by a bitch to her own newborn puppies and can be fatal to them, which is why most litters of puppies are dewormed as a matter of course. Roundworm infection can readily be picked up by humans, particularly children. The worms' eggs can stick to a dog's coat and anyone patting or petting the animal may absorb the parasite through a small cut or even by inhalation. T. canis is, of course, most readily acquired on shoes and clothing via a walk in a park where infected dogs have defecated and this has been a major – if belated – incentive for the law-enforced use of pooper-scoopers and dog-owners' obligation to clear up after their pets. What happens after one of this parasite's eggs has been swallowed is helpfully described by Wikipedia:

> Dogs ingest infectious eggs, allowing the eggs to hatch and the larval form of the parasite to penetrate through the gut wall. In dogs under three months of age, the larvae hatch in the small intestine, get into the bloodstream, migrate through the liver, and enter the lungs. Once in the lungs, the larvae crawl up the trachea. The larvae are then coughed up and swallowed, leading back down to the small intestine, where they mature to adulthood. This process is called tracheal migration. In dogs older than three months the larvae hatch in the small intestine and enter the bloodstream, where they are carried to somatic sites throughout the body (muscles, kidney, mammary glands, etc.) where they become encysted second stage larvae. This process is called somatic migration.

It is for this reason that many dogs and cats are regularly dewormed, and this and other interventions can in turn have severe environmental consequences. But even worse than being disease vectors, our pets can also become disease reservoirs. By June 2022, when monkeypox had become merely the latest health panic, a British epidemiologist warned that the virus – which is closely related to smallpox – could be passed on to pets. These could then become permanent reservoirs of infection exactly as animals in Africa did before passing it on to people.

Even without parasites, pharmaceutical residues or viruses, canine excrement can be environmentally damaging by adding excess nitrogen and phosphorus to the ecosystem. Researchers at Ghent University estimated that in 2020 and 2021, in four nature reserves near Ghent, dogs brought an extra five kilos of phosphorus per hectare and eleven kilograms of nitrogen per hectare each year, leading to loss of plant biodiversity and habitat degradation.[15] Indeed, there is growing evidence that dogs have a huge ecological impact. They are now the most common carnivore on Earth, whether feral or tame, with their global number currently assessed at 1 billion. In the UK alone there are some 13 million pet dogs. Noting how often indulgent owners who take their dogs for a run along the seashore fail to prevent them chivvying nesting seabirds, Aisling Irwin in *New Scientist* of 27 April 2022 wrote, 'At a time when nature is under pressure like never before [...] dogs, whether free-roaming or home-based, are killing, eating, terrifying and competing with other animals. They pollute watercourses, over-fertilise soils and endanger plants.'

ANIMAL MEDICATION

If people often dispose of their own unused medications carelessly, many are even less careful about their pets' old pharmaceuticals. Much of the fallout from pets' anthelmintic (anti-worm) medication remains uninvestigated and hence unquantified. Anti-flea medication is another matter. Up to 80 per cent of the 13 million dogs and 11 million cats in the UK get regular treatment for fleas, whether needed or not. One of the commonest drugs for this is fipronil: a neurotoxin deadly to insects (especially bees), crustacea, fish, plankton and rabbits. Much the same is true of another common flea powder containing imidacloprid, a neonicotinoid insecticide that is indeed fatal to fleas but also to moths, butterflies, bees, birds and much else. It was recently estimated that a single flea treatment of a medium-sized dog with imidacloprid contains enough pesticide to kill 60 million bees. Both these chemicals are now found in most of the UK's rivers, poisoning aquatic insects and starving fish and other aquatic animals dependent on them for food. It may be great fun to throw sticks into a river for a frisky dog to retrieve, but if the animal has been recently deflea'ed with a commercial powder its fun could well prove fatal to a host of other creatures downstream. 'In twenty sampled [UK] rivers, from the Test on the south coast to the Eden in Cumbria [...] fipronil was found in 98 per cent of samples, and the average level of its highly toxic breakdown product fipronil sulfone was thirty-eight times above the recommended environmental safety limit.'[16]

A thoroughgoing study has yet to be made of the full environmental consequences of current animal (or indeed human) pharmaceuticals, although occasionally unexpected – even faintly lurid – cases do get briefly highlighted. These famously include the hormones commonly found in birth control pills that inevitably find their way into rivers, lakes and the sea where they can upset the hormonal balance of, especially, fish. Oestrogen in a Swedish lake led to the eradication of an entire fish population. For that matter, heavy meat eaters of both sexes also excrete quantities of oestrogen because beef cattle, like many farm animals, naturally produce large quantities of the hormone which is anyway present in cows' milk. The horse-wormer ivermectin hit the headlines in early 2021 when it was touted on social media in the US as a 'cure' for Covid. As an antiparasitic it has long been in use, both in humans and animals. It is in extensive and damaging use by salmon farms against sea lice infestation and is highly toxic to invertebrates.

Of greater potential danger to humans is the overuse of antibiotics and hormones in animals, particularly in farm animals. Three-quarters of all the antibiotics currently sold in Europe are not for humans but go to treat farm animals, with an estimated 58 per cent excreted straight into the environment and finding its way into watercourses. This widespread use of antimicrobial medicines on farm animals is one of the leading causes of the rise of antibiotic-resistant superbugs. The practice is hardly confined to cows and pigs. The World Health Organisation (WHO) recently reported that in some countries as much as 80 per cent of all antibiotic use is on animals, including pets. This has led directly to increased resistance to various bugs in humans that in turn leads despairing doctors to reach for the strongest last-resort medicines in the hopes of saving a patient. A range of antibiotic-resistant bacteria has now been traced to raw dog food. According to a press release for a study presented at the European Congress of Clinical Microbiology & Infectious Diseases in July 2021, scientists from Porto University said they had analysed fifty-five samples of dog food from twenty-five brands, including fourteen raw frozen types. The researchers were looking for enterococci bacteria and found that all the raw dog food samples contained antibiotic-resistant enterococci, including bacteria resistant to the last-resort antibiotic linezolid. Genetic sequencing revealed that some of these antibiotic-resistant bacteria in the raw dog food were the same as those found in hospital patients in the UK, Germany and the Netherlands. 'The close contact of humans with dogs and the commercialisation of the studied brands in different countries poses an international public health risk,' said researcher Ana Freitas. 'European authorities must raise awareness about the potential health risks when feeding raw diets to pets, and the manufacture of dog food, including ingredient selection and hygiene practices, must be reviewed.' She added that dog owners should wash their hands after handling pet food and disposing of faeces.[17]

In short, the mere act of giving a man's best friend its dinner should probably earn that man a good soak in a scalding bath of Dettol. All dogs are potentially plague dogs.

INFANTILISM

To stroll the aisles of a large pet store is to recognise how thorough and all-encompassing the infantilising of pets has become in recent decades. In some magical and faintly disgusting way they have ceased to be animals. One might almost be in a branch of Mothercare with the vast range of foods, playthings, objects to help with teething and 'Doggy Dental' products to combat plaque. There are special rugs, cosy sheepskin igloo-cradles for pets to crawl into (the family dog, within living memory, normally slept outside in a reasonably weatherproof kennel); harnesses, leashes and collars, fluffy bootees, elaborate playpens (for mice or hamsters); baubles, bangles and beads of every description.

Then there is the grooming equipment, particularly for dogs. This includes special aprons, nail clippers and nail care products, blades, brushes, driers, scissors, shampoos, conditioners... 'In order to keep your furry friend happy and healthy,' claims one website, 'pet owners must regularly groom their dogs.' Must they? Wolves and dogs in the wild seem to get by perfectly well without being regularly shampooed and conditioned, or having miniature bow ties and multicoloured ribbons attached to their fur. Moreover, the shampoos and conditioners often contain environmentally harmful chemicals such as the already-mentioned flea treatments. The fashion for people to 'cuteify' their pets – so ludicrously prevalent in the United States – ultimately says far more about their owners than it does about the wretched animals they keep captive in city apartments. It is hard to know whether cats are psychologically damaged by being surgically declawed, or dogs by being declawed and debarked and having their ears fashionably cropped. However, it is diverting to imagine a Saki-style short story in which one day some surgically mutilated pets take revenge by spontaneously ganging up on their owner in her New York duplex. Using their expensively cared-for teeth alone, they dismantle her all over the bedroom carpet for her husband (with the heart condition) to find on his return.

A further adjunct of the infantilising pet market that is itself very big business is the pet toy industry. By 2017 this niche market was worth over $1 billion in the US alone. According to the research director for Packaged Facts, a gentleman named David Sprinkle, 'Pet humanisation and "pets as family" trends play pivotal roles in growth within the toy industry.' 'The increasing focus on pet health and wellness will enable the UK to register year-on-year growth at 2.4% in 2021,' claims Future Market Insights. Without wishing to hear how anybody distinguishes 'pet health' from 'wellness', we can cut to the chase and list the sort of toys that responsible 'pet parents' [sic] lavish on their furry children:

> Pet parents have expressed a desire to use more sustainable products that are safe for the environment as well as for pets. Pet toy companies are increasingly opting for sustainable materials for manufacturing pet toys to attract the attention of environment-conscious pet parents. Currently, rubber accounts for the majority of pet toys produced in the market.

Who knew about so many pets' rubber fetish? But the industry's awareness of other possibilities is growing.

> Some of the leading brands offering pet toys in the market have now expanded their portfolios to offer natural pet toys, which might include play balls made with natural environmental cotton and straw rope, and sisal toys or interactive toys made of rope. These natural products are increasingly promoted as chemical-free, nontoxic, sustainable, and biodegradable products which are now gaining popularity among pet owners.[18]

Quite what 'natural environmental cotton' could possibly be is anyone's guess. A plant able to grow without an environment would be unnatural to a degree. In any case, whether made of cotton, sisal, rope or rubber, this global tonnage of toys for animals to play with – or more likely to reassure their owners that the captivity they are imposing can be ecstatic fun – carries with it a considerable environmental footprint in the usual terms of raw materials, processing, packaging, shipping and marketing.

Thus it is that ordinary parents and 'pet parents' gradually become indistinguishable. Early in the New Year of 2022 Pope Francis irritated millions by referring to this as a form of selfishness. 'We see that some people do not want to have a child but they have dogs and cats to take the place of children. You may laugh, but it is a reality.' The more this substitution becomes entrenched, the closer pets also approach human children medically ('pet humanisation' in David Sprinkle's phrase). The sophistication of current research into animals' ailments both physical and mental can even bring a degree of overlap with humans in terms of medication and therapy. The more intensive this trend becomes, the greater the likelihood that new and ingenious molecules constructed by both Big Pharma and Pet Pharma will have unintended consequences for what is left of the 'natural' world. Little enough work has been done on the effects of the excreted by-products of most human drugs and medicines, and the same is true of veterinary pharmaceuticals. After all, these present just as much of a potential threat, covering as they do a wide spectrum of the animal kingdom including pigs, cattle, poultry, fish, felines, canids, lagomorphs (rabbits), birds and rodents (mice, rats and hamsters).

Many people yearn to be loved; and millennia of selective dog-breeding have resulted in canine genes that can produce the heartwarming illusion of absolute devotion. Cats, to their eternal credit, remain aloof and have proved immune to this dyadic conspiracy. It is the infantilising of animals by their surrogate 'parents' that makes one despair of humans ever attaining a relationship with the 'natural' world that is both intelligent and healthy. There is no reason to suppose that animals in the wild ever suffer from boredom; but the plethora of pet toys suggests that many owners must, at whatever level of consciousness, suspect their pets do indeed get bored when imprisoned in an empty house or apartment for long hours while their owners are out. Dogs in the wild are free-ranging. Even the most revolting insect-like lap dog breeds are genetic perversions of wolves, hard as it is to imagine. Maybe in their sleep they experience a mysterious vestigial impulse to run with a pack and howl at the moon; but then it's back to being shampooed and beribboned. To hear people babbling brainless baby-talk to their captive animals is just another reason to despair of the human race. We range in attitude between not wanting to know about the invisible herds and flocks driven 24/7 through blood-drenched killing machines on our own and our pets' behalf, and misreading very stupid animals as human because they can learn to ape winsomeness in exchange for reward, even as they endearingly piss on the carpet.

Gardening

If the function of pets is partly to show your cuddly relationship with the animal world, then gardening is the bright banner on the metaphorical flagpole outside a house that proclaims its owner to be in touch with the soil. Because it implies the excellent ideal of self-sufficiency, gardening has no connection whatever with farming in its modern industrial form. (Modern agriculture has much the same critical impact on the global environment as do transport, lighting and heating in terms of greenhouse gas emissions.)[1] Gardening is indeed a healthy, often exhausting, pastime that obliges one to have an ongoing relationship with the natural order of things. It means paying attention to the weather, the seasons, the soil, the coming and going of animal and insect life. A garden can offer a haven of sanity into which harassed householders can retreat from the world for an hour or two, 'Annihilating all that's made / To a green thought in a green shade' as Andrew Marvell famously put it. When the rather more camp twentieth-century English writer and gardener Beverley Nichols was sneeringly asked by a macho inquirer what he proposed to do if nuclear war broke out, Nichols memorably replied that he would 'shelter behind ramparts of sweet peas'. Witty and snobby, Nichols's views make his gardening books a pleasure in themselves. Among other flowers he cultivated cyclamens – 'those vulgar, obvious plants' – even though he associated them with 'tiresome women who live in flats with electric stoves and indigestion and a Pekingese snoring in the scullery'. He well understood that nuclear war was the very least of reasons for creating a green asylum inside the boundary of one's garden fence or in an allotment.

Humans have, of course, grown their own food ever since hunter-gathering became unfashionable. Britain has long considered itself a nation of gardeners, and for much of the last two centuries the pursuit has had surprisingly deep political roots. After the 1845 General Enclosures Act had dispossessed so many country folk of so much formerly common land, parliament tried to ward off any revolutionary uprisings in sympathy with those already sweeping Continental Europe in the late 1840s by leaning heavily on local authorities to provide allotments in which people could grow their own food. Eventually this was followed in 1908 by the Small Holdings and Allotments Act. The standard allotment was fixed at 300 square yards and carried a small annual rent.

Both world wars were to leave their mark on British gardening. The enemy's strategy of using submarines to blockade island Britain led to an official policy of promoting self-sufficiency and the number of allotments quickly grew. In the First World War local authorities gained the power to confiscate land for the duration of hostilities in order to grow food, and by 1920 there were nearly 1.5 million allotments in Britain that would go on to serve Britons very well during the Great Depression of the 1930s. In the Second World War this number increased to 1.75 million. In addition, the Ministry of Agriculture's Dig for Victory campaign led to considerable areas of the parks in central London and elsewhere being converted to kitchen gardens. The sight of the Albert Memorial and the Royal Albert Hall looking out over rows of potatoes and runner beans in Kensington Gardens soon became unremarkable. So also was a wartime recipe popularised by the then-Minister for Food, Lord Woolton. Woolton Pie was essentially a stew of whatever vegetables fell to hand, topped off with piecrust. It achieved notoriety rather than popularity, being seen as a vegetarian travesty of the traditional steak-and-kidney pie with its rich, meaty gravy.

Meanwhile, the invention of radio meant that gardening programmes were already popular. From 1934 until his death in 1945 a BBC broadcaster named Cecil Henry Middleton had his own weekly radio programme, *In Your Garden*, that had a peak wartime audience of 3.5 million. Mr Middleton, as he was popularly known, was merely the first in a stream of British celebrity gardeners. In the early days of television he was succeeded by Percy Thrower and from then on what already amounted to a national hobby grew steadily, catered for by an increasing number of magazines, radio and TV programmes and – more recently – websites and blogs. Even so, once the post-Second World War austerity had diminished and with the opening in the 1960s of American-style supermarkets in the UK that offered one-stop shopping for all kinds of foods and household hardware, allotment gardening lost many of its adherents. It was obviously so much easier to shop from a wide selection of vegetables and meat all under one roof than it was to dig an allotment, plant potatoes and beans and trust to the weather. Councils up and down the country took advantage of this reduced interest in allotments by quietly selling off for lucrative development land that so recently had been a vital means for the nation's survival in two World Wars.

But social trends never stand still and in the mid- to late-1970s the BBC sitcom *The Good Life* reflected a widespread dissatisfaction with 'the rat-race' and a desire to return to something like self-sufficiency. This prompted a renewed interest in allotments and many took up gardening again, no doubt hoping they wouldn't have to share the experience with a neighbouring couple as gratingly winsome as that played by Richard Briers and Felicity Kendal. Later, in the 1990s, people became conscious of 'food miles' and the long journeys so much of the fruit and vegetables they bought had made before reaching their shopping basket. It was hard to believe that broad beans coming from Italy were 'fresh-picked' as the packet claimed. Fresh when they were picked, no doubt, but steadily losing freshness ever since. Consumers worry about air miles in terms of carbon as well; but they also worry about unwanted and potentially dangerous chemicals hitching a lift into their bodies on the food they buy. In short, back to the allotment where in theory it is possible to have more control over what goes into the soil and into the fruit and vegetables grown there.

Other gardeners concentrate on flowers, for what must be a variety of reasons ranging from the varied aesthetics of Andrew Marvell and Beverley Nichols to the rather more competitive motives that incline them towards show gardens and the awesome social pressures of living in a Best Kept Village. An elderly friend delights in the annual deputations she receives from village worthies begging her to get rid of the wheelless and rusting hulk of a Triumph Mayflower barely visible to one side of her large front garden. They tell her it 'lets down' the village and will yet again strike a fatal blow to its hopes of becoming Best Kept. My friend rightly protests that when the lilac is out the car can't be seen from the front gate. What is more, the now prettily overgrown carcase has long been colonised by badgers. Can these mostly ex-townee 'villagers' really be asking her to destroy the badgers' sett beneath the chassis? Are they aware that this is a protected species (except when being culled by DEFRA marksmen)? And year after year the deputation leaves, muttering darkly.

PESTICIDES AND FERTILISERS

But whether the hope is to grow blooms to Chelsea Flower Show standards or provide homegrown fruit and vegetables, the enduring British craze for gardening has unfortunately come at a high environmental price, nearly all of it in the last seven decades or so. Most of this is down to the various chemical products designed to aid the gardener that have proliferated over those years.

The end of the Second World War also marked the moment when some significant new pesticides and fertilisers began to come on the market. The most notorious of these was undoubtedly DDT, an organochlorine that was the first of the modern synthetic insecticides. It was developed during the war principally to protect troops from insect-borne diseases such as malaria and typhus and it would soon become used in enormous quantities as a general insecticide, and not just by the military or in agriculture. In the United States it was widely and indiscriminately sprayed from the air in huge volumes over towns, farmland and forests. This was evidence of a largely transatlantic post-war enthusiasm that could only ever see glorious advantage in any 'breakthrough' scientific discovery. Fifty years earlier a similarly foolish and ignorant optimism had surrounded Marie Curie's discovery of radium. That famously led to radium being enthusiastically incorporated in toys, chocolate, watches, toothpaste, shampoo, suppositories and a host of other medicaments. For forty-odd years, radium was hailed as the futuristic solution to a host of problems. The awful side-effects of the atomic bombs the USAAF dropped on Hiroshima and Nagasaki in 1945 finally confirmed the dangerous aspects of this supposed panacea.

Likewise, by the late 1950s it had become evident that DDT was indiscriminately killing all insects, the essential as well as the harmful. Moreover, as the American marine biologist Rachel Carson revealed in her 1962 book *Silent Spring*, DDT's organochlorine molecule is exceedingly long-lived and all-pervasive in nature, to the extent that it was being taken up by seabirds whose populations began collapsing dramatically. Carson discovered this was because the eggshells of birds that had ingested DDT were too thin and their eggs disintegrated. As soon as her book was published this heroic scientist was vilified and pilloried in the United States by the manufacturers of DDT, by farmers, by politicians, and by just about everyone else who had fallen beneath the spell of American superpower optimism. Technology of all kinds was the USA's route to the stars; it was not supposed to have drawbacks. Carson was publicly accused of the worst crimes Americans could imagine: of being a communist, of being a traitor, of being a lesbian. Undercover agents spied on her. Chemical and agricultural companies paid large sums to their own scientists to come up with evidence casting doubt on her scientific competence. Unfortunately for them she was right and it wasn't long before molecules of DDT were found in practically any soil sample taken anywhere on Earth, as well as in the ocean deeps. Today, long after DDT has been banned for all but the most exceptional use as an antimalarial in a handful of tropical countries, its all-pervasive molecules are still to be found everywhere. They now jostle with microplastics and polymers in the longevity stakes as well as in our bloodstream. And nor is DDT alone as a persistent organochlorine of that period. Another was chlordane, an equally effective insecticide that was lethal to fish. It was banned in the UK as long ago as 1981. It takes years to degrade in soil and accumulates in animals and humans, and even today its metabolite oxychlordane can be found in the blood of people born long after the substance was banned.

The last paragraph of Carson's famous book began with a sentence that could stand as a credo for all environmentalists everywhere, for all farmers and also for all gardeners: 'The "control of nature" is a phrase conceived in arrogance, born of the Neanderthal age of biology and philosophy, when it was supposed that nature exists for the convenience of man.' Notwithstanding this, the world's principal business model, capitalism, does indeed treat nature as existing for human and economic convenience. In the last seventy years a constant stream of insecticides, fungicides, herbicides and fertilisers has poured from chemical companies, all of them designed precisely to control nature by attempting to solve agricultural and gardening problems worldwide. Very many of these molecules have ended by being phased out or even banned as their deleterious side-effects became fully apparent. Notorious Goliath-and-David battles have been fought between giant corporations and individuals who claim their health has been harmed by contact with various of these chemicals.

INDUSTRIAL FARMING LEAVES ITS MARK ON GARDENING

Monsanto's glyphosate-based weedkiller, Roundup, bought by Bayer in 2018, has shown itself of often critical importance to crop-growers worldwide and is now widely manufactured, with 30 per cent of global exports coming from China. However, glyphosate has also been increasingly suspected of serious collateral damage, especially to wildlife and human health. It has been the subject of over 100,000 lawsuits in the US alone. These were brought by former Roundup users claiming that the weedkiller had caused them to develop such things as non-Hodgkin lymphoma and eye damage. Bayer agreed to settle their claims by paying out some $14 billion and further promised to stop selling glyphosate in the US by 2023.[2] The EU's own current approval for Roundup, the world's most widely used herbicide, expires in December 2022 and may not be renewed. Austria banned it in 2019.

Roundup's insect-killing counterpart, described by Wikipedia as 'the most widely used insecticide in the world' is imidacloprid. This is one of a group of chemicals known as the neonicotinoids that has come under increasing attack and in 2013 the EU banned three of them. As mentioned in this book's chapter on pets, imidacloprid has been – and still is – a common active ingredient in flea powders for cats and dogs. All three of the most widely used neonicotinoids are deadly to honeybees, aquatic invertebrates and birds. A study for American Bird Conservancy in 2013 found that 'a single corn kernel coated with a neonicotinoid can kill a songbird'.

Many of these industrial chemicals became progressively available to farmers and British gardeners from the 1950s onwards. They have unquestionably played a major role in the slaughter of insects and other wildlife that makes today's gardens barren by comparison with how they were half a century or even twenty-five years ago. In his 2022 book *Regenesis*, George Monbiot described pesticide use in British agriculture as 'Insectageddon'. Nor is this slaughter restricted to Britain. In only the last twenty years 70–80 per cent of Denmark's insects have vanished. I was lucky enough to grow up in an acre of Kentish suburban garden on the edge of London in the late 1940s and early 1950s and as a boyhood collector and breeder of butterflies and moths I could easily find thirty different species without setting foot outside the front gate. Overnight the garden was also full of hedgehogs that would often get entangled in tennis netting and need freeing in the morning. We would regularly see the odd badger or not-yet-urbanised fox after our hens. Three magnificent old elms housed all sorts of creatures including families of owls. Britain's elms were mostly wiped out in the 1970s by an imported fungal disease spread by elm bark beetles. Their loss has changed the landscape forever and many of the fauna and insects once dependent on them have now died out in suburbia. But because gardening is still passionately practiced there, a good deal of this mortality of wildlife must be laid at the door of the various pesticides, weedkillers and growth hormones with which gardeners have innocently anointed their plants and soils over the last half-century in an attempt to 'get back' to their own vision of Nature. In many cases that vision has included exotica imported from overseas; and with these plants have come many non-native diseases and insects. Among these was Dutch elm disease itself, imported on infected timber.

IMPORTED PLANTS

Ever since at least the nineteenth century, the fashion for importing exotic plants and timber has had awful consequences, not only for gardens but for the landscape itself. Today we are all too familiar with plagues of invasive species such as rhododendrons in Scotland, Himalayan balsam and Japanese knotweed everywhere else. This latter in particular has proved exceedingly difficult to eradicate and may be so invasive it can lead to a property's loss of value. When asked by despairing homeowners what they can do against their infestations of Japanese knotweed, BBC Two's *Gardeners' World* has been reduced to recommending spraying with Roundup or similar glyphosate weedkiller.

Nor is the damage caused by the fashion for exotic houseplants confined to importing countries. It has led to ruinous poaching in tropical forests, such as for Philippine pitcher plants, and equally in temperate woodland for lady's slipper orchids. Such imported species often bring foreign insect eggs and fungi with them; but just as importantly, the removal of any wild species of plant or animal leaves behind it a gap in local diversity, a potential break in the chain of pollination and reproduction. Apart from that immediate damage, the ruthless trade in exotic plants from the tropics requires the extensive use of heating for transport and greenhouses before the poor alien can go on display in someone's centrally heated living room, often thousands of miles from its natural habitat. The imported plant trade is the botanical equivalent of the cowboy trade in tropical aquarium fish.

PESTICIDES

In the post-war years there was at first little in the way of commercially available and effective pesticides. Our gardener collected his fag-ends in a big sweet-jar of water where they gradually unravelled to produce a noisome brown liquid he would spray on the roses against aphids and mildew. This nicotine-containing drench was reasonably effective but short-lived. Before the Second World War and even before the First he might have sprayed a solution of whale oil, which could not have much enhanced the blooms' scent. (Today's equivalent might well be a solution of Neem oil.) It should be pointed out that there are at least three naturally occurring insecticides that have been in use for centuries: nicotine, pyrethrin and derris. Gardeners have known about tobacco's insecticidal properties for at least 300 years. Pyrethrin derives from chrysanthemum flowers which the Chinese were using as an insecticide 2,000 years ago. And a widespread tropical vine, Derris elliptica (also known as tuba) contains a chemical called rotenone and was historically used by indigenous peoples in South America as a fish killer and in South-East Asia against mosquitoes. However, just because a substance occurs naturally it doesn't mean it's harmless, as anyone who mistakes a death cap for an ordinary field mushroom soon finds out. Once modern chemists had analysed these plants and isolated their active substances they set about producing derivatives. Nicotine's problem was always that it degrades quickly, and farmers and gardeners wanted something longer-lasting and more selective. In due course industrial chemists' molecular juggling spawned a family of new nicotinoids duly known as the neonicotinoids. Pyrethrin likewise produced a number of pyrethroids with differing characteristics; and the tuba vine yielded derris powder and a family of chemicals known as rotenoids. All these naturally derived substances have drawbacks as well as advantages.

It wasn't long before nicotinoid concentrates appeared on the market for gardeners to dilute according to what they wanted to kill. In the US there was Black Leaf 40 which contained 40 per cent nicotine sulphate and was so fiercely poisonous to man and beast (let alone to aphid) it was soon largely discontinued. Several neonicotinoids have been banned, the latest being thiamethoxam. This was outlawed across Europe in 2018 after a series of studies found it damaged bees. However, British Sugar applied for an emergency exemption and in early March 2022 it was announced that the conditions for exemption had been met. Thiamethoxam is sprayed on sugar beet to stop a disease called 'virus yellow' spread by aphids and which in 2020 (according to the UK government) destroyed 25 per cent of the UK sugar beet crop. Since about two-thirds of the UK's sugar now comes from homegrown sugar beet this is yet another case of the constant tension or trade-off between economic advantage and environmental damage.

As for Derris Dust, this was eventually banned in the UK, but only as late as 2009. Some half a million British gardeners are members of the Royal Horticultural Society and a glance at its website will provide a list of at least twenty chemicals that were commonly used in British gardens until withdrawn or banned in the last twenty years. Apart from Derris Dust they include malathion, aldrin, dieldrin, Bordeaux mixture and formerly widely used herbicides like paraquat. This notoriously killed children who mistook it for lemonade and has been linked to Parkinson's disease. Needless to say, because products containing condemned chemicals have been banned from sale it does not mean that gardeners may not still have supplies of them, whether hoarded in a tool shed or obtained via the Internet.

Anyone who has read the earlier chapter on pets will recognise an overlap between these two worlds. As just noted, many of the insecticides used to treat fleas in cats and dogs are the same as those used in gardening and agriculture. Many of the more serious animal treatments are prescription drugs and in theory only obtainable from a qualified vet; but of course regulations vary widely the world over and animal pharmaceuticals – like their human counterparts – are often easily available on the Internet. Pyrethroids like permethrin, cypermethrin and tetramethrin occur over and over again for both animal and garden use. Tetramethrin is merely one of 320 pesticides banned for sale in the EU but is readily available in the UK as the active ingredient in Pest Shield household flea spray. All pyrethroids are highly toxic to bees, aquatic insects and many varieties of fish and birds. Another group of chemicals derives from bacteria found in rotting sugarcane: spinosad and spinetoram. Spinetoram controls and kills worms and caterpillars as well as fleas and some beetles. It is sprayed on crops as well as used to kill fleas in cats under the brand name Cheristin. Predictably, it is also toxic to insects, birds, freshwater invertebrates, aquatic plants and mammals. As with other chemicals the forms it takes as it degrades in the environment can be even more toxic than the original substance.

All this is enough to indicate the crossover between animal care and gardening of various groups of chemicals. Manufacturers in both fields are of course well aware of the extreme difficulty of finding a magic bullet that is specific to a particular pest or virus and harmless to everything else. Great progress has been made since the days of DDT with its noxious blanket effects on the insect and animal kingdoms, and products are constantly being reviewed. For example, the old remedies against slugs and snails most often included metaldehyde, which is highly effective against molluscs. Unfortunately the coloured pellets are also lethal to cats and dogs as well as to toddlers unless they get very prompt treatment. In the UK, at least, metaldehyde is now banned for outdoor (as opposed to greenhouse) use.

GARDEN CENTRES

As well as in hardware stores, such modern aids to gardening as fertilisers and insecticides were once mostly sold in old-fashioned nurseries where at weekends gardeners would also buy their trees, shrubs, plants and seeds. Most nurseries have long since mutated and swollen into those temples of tat known as garden centres. In essence these places are themed supermarkets. Outside there is often the vestige of a nursery with an open-air plot containing larger bushes and saplings for sale. There are also polythene-packaged bags of compost and potting compounds. Some of these still contain peat, which has long been dug up for gardeners despite peat bogs constituting the UK's largest natural carbon sink, a single hectare of which traps as much as does the same area of tropical rainforest. Digging up peat not only releases CO_2 but destroys the habitat of several rare wildlife species. On top of that, peatlands help to reduce flooding. Back in 2011 the UK government set a voluntary target for compost retailers to end sales of peat by 2020. But 'voluntary targets' have all the efficacy of 'self-policing'. Between 2011 and 2019 peat use fell by only 25 per cent and in 2020 it promptly increased by 9 per cent as Covid lockdowns boosted gardening as a hobby.[3] In August 2022 the government confirmed the sale of peat will be banned to gardeners (who use 70 per cent of it) from 2024 and to the horticulture industry 'by 2028'. The whole issue of peat bogs being dug up (mainly in Ireland) when there are plenty of substitutes available such as wood fibre and bark has long been an example of nature being trashed in order to enhance somebody's garden hundreds of miles away. Besides, whatever happened to the idea of a compost heap and well-rotted humus?

Most garden centres have a vaguely 'greenhouse' indoor area where one can buy smaller plants and seedlings. We venture in, trying not to notice the garden furniture made out of 'teak from sustainable sources' – a phrase that sounds as though it deserves a cynical snort and a muttered 'You wish.' Much of the rest of the indoor space is given over to accessories with a loose connection to gardens: tools, pots, bird tables, deck chairs, sunshades, hose fitments, electric insect zappers, rubber boots and the rest. This area normally shades into one of pet foods and pet accessories such as wicker dog baskets and those padded posts that by long tradition cats shun in favour of sharpening their claws on the sofa. Beyond this lies the realm of unashamed tat: a horrid world of scented candles, aromatherapy vials, wind chimes, lanterns, CDs of breaking waves and birdsong, sweets, toys, joss sticks, floral writing paper, 'gifts', model animals from Germany and brightly packaged stuff to attract children that includes yet more sweets. Most garden centres also boast an obligatory café. The dread word 'organic' litters a handwritten menu that usually offers bowls of flatulence-inducing soup made from compost and beans; a vegan quiche containing what looks like bracken; Earth-mother flapjacks; carrot cake; and apart from unspeakable caffeine-free coffee apparently made from freshly toasted twigs there are infinite varieties of tea. These may well include nettle tea, chakra tea, balancing tea and, on one memorable occasion, knock-me-up fertility tea.

The underlying theme of the entire place is middle-class shopping combined with an outing for pacifying the kids at weekends. Dad wheels his bags of fertiliser and Mum's tray of seedlings out to the family's SUV as the kids pile into the back, mouths smeared with brownies and their new stuffed animals bulging with natural, non-allergenic wool. If this all sounds a disaffected way of describing garden centres, it is. But it's also a way of pointing up how much of families' free time is filled by just buying *stuff*; and that shops selling an artfully packaged atmosphere vaguely redolent of gardening and outdoor activity do so with a pungent whiff of utterly bogus virtue. It is merely yet more retailing, naked but for the green wellies and the Richard Briers grin.

DAMAGE IN GENERAL

Noxious chemicals are only one form of environmental damage that gardeners, all unawares, can cause. One problem with allotments, for instance, is that most are in demand and seldom if ever allowed to go fallow. That means if an allotment has been constantly used for sixty years to grow fruit and vegetables, a well-established population of various pests will also have taken firm root and may be extremely difficult to deal with and harder still to eradicate. On the other hand a garden that is allowed to become something of a jungle may not yield much in the way of fruit and vegetables or even flowers, but it will be a refuge for wildlife. Unfortunately, it will probably also become a public lavatory for all the neighbouring dogs, cats and foxes, and the parasites of Toxocara canis and toxoplasmosis are almost certain to be in the soil and ready to be tracked back into the home.

At the opposite end of the scale there is the over-gardening that leads to a loss of plant (and weed) diversity and consequently to a loss of habitat for useful insects. Lawns shaved to within a centimetre of their lives and regularly anointed with various moss-killers and other chemicals are one thing; but grubbing up every last nettle beside the tool shed is another. Several caterpillar species love nettles, as do the chefs in the garden centre's café. And zealous gardeners often unintentionally destroy birds' nests when clearing undergrowth in spring, having not recognised or even noticed them. The UK now has a bare 53 per cent of its biodiversity left and a quarter of native British mammals are now at risk of extinction. By the same token pet owners ought to think about their tame animals, especially the cats that cut such a swathe through wildlife.

One particular article, paradoxically often for sale in garden centres, sums up an entire mindset that betrays a rooted refusal to tolerate nature and an equally rooted determination to shape it. This is the electric fly zapper that indiscriminately lures insects with ultraviolet light and kills them with a sharp crackling of sparks and the smell of burnt chitin. Since these traps are generally switched on at night they kill a host of harmless nocturnal insects, further endangering countless species of moths and micromoths. It is not quite clear what the purpose of this mindless slaughter is, other than (in Rachel Carson's phrase) an arrogant belief that 'nature exists for the convenience of man'. A related phobia surrounds wasps and even to some extent bees, both of which vital insects are generally harmless unless provoked and are anyway hardly life-threatening except in the rare case of people who go into anaphylactic shock when stung, most of whom will anyway already know this and carry an EpiPen. It is grossly disproportionate that everyone else should wish to wipe out any insect that could conceivably sting them. Hornets in particular are, despite the bad press they get, beautiful and mainly peaceable creatures – as well as being officially protected in certain European countries (but not yet in Britain). It seems grotesque that they and wasps should be deemed to merit a barrage of lethal chemicals aimed at them in the form of wasp and hornet bait even when they are minding their own business. For at least the past half-century school biology lessons have explained the idea of a food chain; yet one's neighbours nevertheless persist in buying an electric insect zapper and later, over the garden fence, wonder why in recent years 'you hardly ever see swifts and swallows and martins like in the old days'.

As for the widely available domestic fly sprays, many contain permethrin and tetramethrin. They may have the phrase 'Indoor Use' on their cans but there is nothing to stop a cloud of aerosol drifting out of an open window. Their molecules travel and seem nearly indestructible. Once outside, their insecticidal activity continues, especially in water. Worse still are things like Provanto's Ultimate Bug Killer which contains CIT/MIT. This isothiazolone biocide was introduced by Dow Chemicals in the early 1980s and because of its effects on humans as well as the environment it has been called the new DDT. It has been added to all sorts of products where bacteria can grow, including jet fuel and shampoo, despite it being a serious allergen. It is also extremely poisonous to all sorts of microorganisms, including the vital soil organisms on which gardening and agriculture depend. It is persistent in the sea bed for six months. In short, nobody using household fly sprays can possibly call themselves 'environmentally aware'.

As a codicil we might add that in an ironic reversal, even scrupulous gardening has recently become blighted by environmentalism. Perhaps the saddest victim of this supposedly Earth-saving activity is the humble bonfire. Until recently lighting a bonfire was one of gardening's most treasured rituals. The smoke drifted up in clouds of autumnal incense: a nostalgic scent from childhood. Leaning on a pitchfork and watching the flames gradually take hold was deeply satisfying. Occasionally raking unburnt outliers into the fire led one to recognise withered old friends like the Michaelmas daisies that not long ago bloomed in their prime. In this way a pleasing melancholy of Eheu fugaces perfumed one's tritely philosophical musings on the seasons and death. But lately bonfires are much restricted or banned outright. Nowadays garden refuse is binned and emptied by council workers who truck it away to feed power stations with biomass. This may make ecological sense but it can leave a gardener feeling it is just one more sacrifice made to the Moloch of electricity generation and the self-righteousness of one's neighbours. Who, incidentally, think nothing of polluting the entire area with the bellowing and exhaust of their two-stroke leaf-blower (another slack-jawed purchase. Whatever happened to the humble, silent, leaf rake?). When not deafening the neighbourhood by day they enjoy sitting out on their patio on autumn evenings beneath a great gas-powered mushroom heater bought – like the moth zapper glowing behind them – at the local garden centre. As the writer Martin Cruz Smith once observed, 'Normal human activity is worse for nature than the greatest nuclear accident in history.'

Sport

A good many sports are increasingly entangled in the gelatinous tendrils of the wellness industry, oddly enough for participants – who are already presumed fit – as well as for spectators. But of course Peak Fitness (which sounds like one of those mountains in the former Soviet Union) is a summit from which many fall to a premature death. However, a vigorous game of tennis or squash, a session spent in a gym, swimming lengths in the local pool or a round of golf are all sensibly viewed as virtuous adjuncts to a 'healthy lifestyle'. Equally strenuous – even downright dangerous – sports such as foxhunting tend not to be mentioned on the mistaken grounds of being purely elitist, as well as registering low on the animal-friendly checklist. However, unlike the unstructured informality of a walk or a bike ride in the country many, if not most, sporting activities also require a specialised site and equipment to which spectators are attracted. The more televisual the activity and consequently the bigger the business, the more the impact of spectators needs to be added to determine any sport's environmental footprint.

MOTORSPORT

In general, motorsport probably attracts the most criticism since it appears to be so triumphantly unGreen. Of all the various categories that include rallying and MotoGP it is probably Formula 1 that looks the most defiantly unreformed. The crowd-pleasing, hi-revving shriek of finely tuned internal combustion engines, together with the shuttling about the planet of the F1 teams with their cars and engineers, has been called a 'carnival of carbon'. Formula 1 is indeed responsible for just over a quarter of a million tons of CO_2 emissions annually, although more than half of this comes from the air, sea and road transport of F1 personnel, promoters and partners. Some of this is already offset with carbon capture technology and tree-planting. The old *Top Gear* team's ethos of treating the natural world as less than the dust beneath their Pirelli radials has mellowed to a more eco-friendly attitude among petrolheads. Jeremy Clarkson has even taken up farming, for pity's sake: less a straw in the wind as an entire haystack. Many top-ranking F1 drivers like Sebastian Vettel, Lewis Hamilton and Fernando Alonso have declared themselves firmly on the side of environmentalism and the need for motorsport to embrace a steadily more hybrid future. Indeed, Hamilton got rid of his private jet on environmental grounds. Yet he and his team still need to fly about the world in order to get to widely scattered Grands Prix in today's tight racing schedule.

The constant mobility of the F1 circus does attract increasing awareness and criticism. But it is the defiantly internal-combustion nature of the cars' engines that makes them an obvious target for the environmentally aware, even though the cars are already powered by 10 per cent 'Green' (i.e. biomass-derived) fuel. However, the one thing the global millions who follow F1 agree on is that the nearly noiseless swishing of Formula E racing cars is nothing like enough to satisfy them – not even close. The sheer sound of highly tuned engines is essential. That high-pitched hysterical howl as the pack leaves the starting grid conveys all the excitement and effort of maximum output, producing a sense of driver and machine performing at their very limit. And since F1 is very big business indeed this requirement is not likely to change any time soon.

Ross Brawn, the Formula 1 managing director and former engineer, puts the matter squarely into perspective. For him it is simply a matter of physics. The present state of technology is nowhere near able to get the necessary energy density to power a Formula 1 car without using fossil fuel. Today's typical F1 car weighs a bare 750 kilos. Its engine generates over a thousand horsepower and it can maintain this output for races lasting a couple of hours or more. Indeed, the car's limits are far more usually set by the life of its tyres. By contrast, a Formula E car representing the pinnacle of electric-powered motorsport typically weighs 900 kilos because of the heavy battery, and generates a mere 300 horsepower. What is more, the power demands of racing are so severe that e-car drivers are obliged to swap their car for a fresh one halfway through a race because the batteries simply cannot hold enough power. Even by using the latest battery technology there is no way e-cars can get close to the extended high performance of F1 cars. To rival their power output alone with today's technology, an e-car's battery would need to weigh some six or seven tons, which would make the car itself an entertaining spectacle but not much use for racing.

Ultimately, of course, the performance rating of any engine comes down to its power-to-weight ratio. We tend to forget that exactly the same calculation applies to all other vehicles. Those who imagine that intercontinental e-airliners will soon be flitting about the world making little more noise than a strange moan are to some extent still inhabiting the world of well over a century ago when cartoonists envisioned cities of the future above which steam-powered aircraft with sails whizzed busily about. Small, short-range e-aircraft may indeed be a possibility within the next decade or so (see the Cars & Aircraft chapter), but nothing capable of crossing oceans cheaply with a couple of hundred passengers. Yet going electric may not always be the answer, especially since so much of the world's electricity supply is still generated by fossil-fuel-fired power stations running on coal, gas or oil. What needs to be developed, according to Brawn, is a sustainable, net-zero-carbon fuel.

For the foreseeable future, therefore, F1 will remain much as it is while the synthetic component of its fuel will steadily increase. (But this, for the reasons outlined in the Cars & Aircraft chapter, may not always be quite the gain it seems.) Even so, Formula 1 itself claims that only 0.7 per cent of an entire season's carbon emissions is from the cars themselves.[1] A hydrogen/air mix could in theory power racing engines while still producing that satisfying internal-combustion sound the fans demand, but the technology capable of making it safe while producing the same or greater power is still far from realisation. Hydrogen is easily obtained by the electrolysis of water, but of course that process itself demands a good deal of power in the form of electricity that in turn must be generated somehow.

In short, the motor racing world awaits a breakthrough. The same applies to MotoGP motorbike racing, as it also does to rallying and – particularly in the United States – to dragster and powerboat racing. In fact, both these last use nitromethane ('nitro') as fuel which can be Greener in origin but produces nitric oxide (NO) and CO_2 in the exhaust. Truck and tractor pulling, with its behemoth machines like bulls pawing the ground and snorting dense clouds of black carbon, will surely survive in the states of middle America for triumphant enclaves of supporters with gun racks in their pickups who anyway believe global warming is a socialist myth.

GOLF

The actual venues of spectator sports come at considerable environmental cost. Football pitches with their undersoil heating; swimming pools with their tens of thousands of gallons of heated and chlorinated water; ice rinks with their power-hungry ice-making machinery; the manicured arenas for tennis, cricket, rugby and horse racing – all demand expensive care and resources. But they probably all dwindle in significance when compared with the construction and maintenance of the world's golf courses.

It would take a thoroughgoing worldwide investigation in order to pin down the cumulative environmental price of maintaining the planetary total of nearly 40,000 golf courses, 78 per cent of which are located in just ten countries: the United States, Japan, Canada, England, Australia, Germany, France, Republic of Korea, Sweden and Scotland. Eighteen-hole courses can occupy anything up to 200 acres of land, the average being about 160 acres, all in need of constant landscape gardening as well as the cosseting and manicuring of the greens. The amount of fertiliser, weedkillers, insecticides, man-hours and above all water required to maintain all these courses must be nearly incalculable. I remember a scandal as long ago as the 1970s when a Portuguese town in the Algarve held angry protests against its community of largely British retirees who, in order to keep their golf course green, were diverting the locals' already feeble water supply.

Such things must be familiar the world over, particularly wherever the US military is stationed for any length of time. The verdant Muroc Lake Golf Course at Edwards Air Force Base in the high and famously dry Mojave Desert is a case in point, not to mention equivalent courses in Iraq, Saudi Arabia and virtually anywhere else that the American armed forces have pitched their tents in the last seventy years. Quite what the locals who daily trudge long distances to fetch battered canisters of water from distant wells and pumps must think on seeing this lifeblood poured back into the ground just so that a handful of people can walk about trying to hit little balls into holes with sticks can only be surmised. But there, the sheer hubris of building entire cities, let alone golf courses, in the middle of a desert (Las Vegas, Palm Springs) is becoming more painfully evident as drought associated with climate change increasingly grips the south-western United States. In June 2021 southern Nevada saw the first municipalities in the US ban the watering of 'non-functional' grass, allowing the strips of grass along sidewalks to die. This is inevitably the shape of things to come, and on a much bigger scale. Yet within twenty miles of Palm Springs there are ninety golf courses: forty-nine private, thirty-seven public and four municipal. It will be interesting to see how well they fare in the years ahead. Their combined water requirement must already infuriate Californian farmers watching their valuable almond orchards shrivel from drought. Phoenix, Arizona – currently billed as The World's Least Sustainable City where before the end of this century temperatures are predicted never to fall below 100°F (37.7° C), day or night – is even now being described as near-uninhabitable in the summer. Despite this there are twenty-six golf courses in Phoenix, plus another seventy-four within twenty miles. There comes a point where a greenskeeper is not Green.

Yet the environmental cost of maintaining golf courses is often only half the story, the other half being the damage caused in the first place by clearing a site for the course to be laid out. This supplied a theme for one of Carl Hiaasen's angry, eco-impassioned satires, *Native Tongue* (1991). Set in South Florida, much of the novel's action concerns the illegal bulldozing of hectares of coastal forest and mangroves for a golf course. That particular course might have been fictional, but its construction was based on the fact of there being 1,250 golf courses in Florida – more than in any other state, and counting.

SPECTATORS

The true carbon footprint of many sports is pushed still higher by their fans' and spectators' attendance. Each year millions around the globe travel considerable distances to follow their team or their hero. They are always there to cheer the Rafael Nadals and Lewis Hamiltons regardless of where they go. In 2015 and quite by chance I encountered five individuals who flew from the UK to Australia simply to watch the Test cricket series of that year. They were part of the well-established 'Barmy Army' of British fans who travel everywhere to support English cricket. There must have been hundreds – even thousands – of others who racked up airmiles for the same event. The Barmy Army began as cricket supporters but has since branched out into Rugby Union and League.

Yet their numbers pale into insignificance when compared with the hordes of fans who fly all over the world to support their football teams worldwide, especially for such soccer bacchanals as the World Cup. Like the universe itself, European club football is in a state of constant expansion. In 2018 European football's governing body UEFA announced the introduction of the Europa Conference League (ECL) in order to ensure 'more matches for more clubs and more associations'. It somehow forgot to mention that this would also generate a great deal more money. Having calculated some of the environmental consequences of this new league, BBC Sport very properly addressed the subject on 21 October 2021. Their bullet points included:

- In the Europa Conference League, 85% of group games are between teams at least 1,000 km apart.
- The average distance travelled per team in the group stage is 5,578 km, near enough the same as flying from London to New York. Across the entire group stage, that would give each person travelling an average footprint of 1,087 kg of CO_2 (more than one tonne).
- Nineteen teams (59%) have to travel more than 5,000 km in total for their group games.
- Forty-two games (44%) involve trips of 2,000 km or more.
- Eleven teams (34%) have to travel more than 1,500 km for each of their away games.
- The Kazakh side Kairat have the longest travelling distance of any team – 11,348km in total: more than

a quarter of the way around the world. A Kairat fan travelling to each of their team's away games would have a footprint of an estimated 2,212 kg of CO_2. That is more than half the world average per person for all activities for an entire year (4,000 kg).

British clubs must be among Europe's worst offenders. Despite inhabiting a comparatively small island, teams nearly always fly to Premier League matches regardless of how short the journey is. In September 2021 Greenpeace criticised Tottenham for taking a twenty-minute flight to Bournemouth. In the following month Manchester United flew to Leicester: barely a hundred miles away with a flying time of about twenty minutes. In October Leeds United took a plane to Norwich for a match that Norwich City had declared would 'champion sustainability practices'. So much for Green pretensions. As the BBC pointed out, '[CO_2] emissions per kilometre flown are known to be significantly worse than for any other form of transport, with short-haul flights the worst emitters, according to the Department for Business, Energy and Industrial Strategy.'[2] France has now banned all short-haul flights for any journey that can be done by train in two and a half hours or less and Spain and Germany may do the same. In October 2021 the UK's Campaign for Better Transport called for a ban on all domestic flights where the journey could be done by train in five hours or less. British commuters might add sarcastically that it is not unknown for a rail journey of a mere fifteen miles to take five hours if, indeed, trains are running at all. It remains to be seen whether British football will change its habits even though it is short-haul flights that do the most damage.

As for spectators and athletes travelling to and from the four-yearly orgies of the Winter and Summer Olympics (not to mention the Paralympics): those must rack up a global carbon debt beyond accurate computation, to add to the huge financial outlay host nations bizarrely compete to shoulder. Nor is the only environmental hazard that of long-distance travel. Things get worse during a pandemic. As a result of the 2020 skiing season the Tyrolean town of Ischgl became the subject of a civil suit alleging that local authorities had been too slow in recognising the resort's role in acting as a Covid 'superspreader'. Despite numbers of people falling suddenly ill, probably as a result of crowded après-ski parties, thousands were blithely allowed to leave the valley unhindered while thousands more arrived with no idea they might be in danger of infection.

High spectator attendance at sports frequently involves people driving to the venue in their own cars. This is especially true of motor racing, many of whose circuits (like Silverstone and the Nürburgring) are sited away from urban areas. In this as in many other popular sports, spectators' cars often queue to get into or out of a track or stadium. This can mean hours of idling engines, particularly in hot weather when the cars' air conditioning is kept running. An army of aficionados follows Formula 1 races as they move from country to country. In 2022 305,000 spectators converged on Zandvoort (normal population around 7,000) to watch the Dutch Grand Prix. The figure for Silverstone was 401,000. The popularity of F1, arguably the most charismatic and lucrative branch of all motorsport, can readily be deduced from its steady growth. In 1950, the very first season of the new Formula 1, there were just seven races. In 2022 there were twenty-three. However, unlike world football it seems unlikely to expand much further since it is limited partly by the severe physical demands made on drivers but principally because of the frantic rebuilding of the cars and engines that takes place between races, to say nothing of getting everything packed up in time for the next leg of the flying circus.

So what of the carbon emissions from one season of Formula 1 cars? These have been calculated at 256,551 tons[3] – a figure that does not include travel by fans, nor that of the small army of international radio and TV reporters' crews. Furthermore, the actual racing cars themselves account for a mere 0.7 per cent of F1's total emissions, which makes much of the criticism levelled at the sport somewhat misplaced. Very much worse are the huge international wingdings of the FIFA World Cups every four years. Tentative calculations were made in advance for the carbon emissions of the 2010 World Cup in South Africa. Those were rated at 2.75 million tons, of which 1,856,589 tons would be racked up by fans flying in from all over the world.[4] In the event the total was slightly less than predicted, but still nearly seven times greater than that of the 2006 World Cup in Germany. Must these massive emissions increase incrementally every four years in pace with the razzmatazz? We shall see. In an instance of transcendent idiocy the 2022 World Cup was bid for and won by Qatar with a promise to be carbon neutral, which can't possibly include the flying-in of the million-plus fans it hopes to attract. Qatar already ranked as the world's highest per capita carbon emitter with 38.82 tons in 2019. Despite its water scarcity and desert environment, Qatar also ranks top in the world on water consumption per capita (557 litres per day) as well as waste generation per capita (1.2 kg per day).

Similar lunacy arranged the 2022 Beijing Olympic Winter Sports in a region of China that each winter has on average less snow than London. The modest requirement was for 1.2 million cubic metres of artificial snow, to generate which needed a lot of electricity and an estimated 49 million gallons of water in an increasingly parched region.[5] Then in October 2022 it was announced that the 2029 Asian Winter Games will be hosted by Saudi Arabia, famous for its alpine ski runs. At least these cases offer the entertaining spectacle of official liars fluent in the language of state-of-the-art Greenwashing.

At the other end of the scale is an example of a sport that ought to rate extremely low in terms of its environmental impact: the 2021 London Duathlon that took place in Richmond Park. It was billed as the world's biggest duathlon – which consists of running and cycling. Hardly a threat to the environment, one would think, even though the competitors themselves inevitably convert a fair amount of atmospheric oxygen into carbon dioxide (with the occasional involuntary blurt of methane). But the event's true footprint was that of the spectators' cars that caused daylong traffic gridlock for miles around beneath rising clouds of exhaust fumes. A month later on a visit to Henley I saw a poster outside St Mary's church, just over the bridge that forms a famous traffic bottleneck. It read, 'Idling is sinful'. A local decoded this for me as not being a reference to personal sloth, as I had hoped, but to drivers leaving their engines running when in a traffic jam. It was good to see the Church for a change scolding sinners rather than bending over backwards to understand them.

The truth is that the more popularly followed sports are not really sports so much as high-revenue spectator industries. Why else would a suburban tennis club suddenly want another thirty-nine grass courts together with a new stadium to seat 8,000 spectators? This is what the All-England Tennis Club at Wimbledon says it needs: additions that will triple the size of the club's existing premises and swallow up Capability Brown's parkland. Applying for planning permission, the club complained that all the other Grand Slams around the world can host their own preliminary rounds while poor Wimbledon has to outsource theirs. But so what? Can it matter where the preliminary rounds are played? Well of course it does. This is quite as much a spectator industry as it is a sport. It's all about money.

E-BIKES

It is obvious that a lot more work needs to be done to determine the true carbon footprint of sports that most people wouldn't consider – or might even deny – have one. Because so many sporting activities are associated with health, leisure or entertainment there is little personal motivation to assess their real cost to the environment. From this it is possible to formulate an unsurprising general law: that people are least able (or most reluctant) to assess the environmental impact of anything they enjoy. Whether it involves actively doing or passively consuming, the pursuit of pleasure is the very last thing most people feel like stigmatising and least of all in an era of epidemics, lockdowns and economic crises. Sugar that pill!

In this context e-biking seems to offer a wider public a plausibly innocent combination of transport and exercise, being less strenuous than unassisted cycling. Currently the popularity of e-bikes in the UK has yet to reach the level it has in Continental countries such as Germany, Austria and the Netherlands, yet it can only be a matter of time before the British market catches up. By 2023 it is projected that the worldwide number of e-bikes will top 300 million. They are already most numerous in China, where the majority are powered by lead-acid batteries that are much cheaper than the Li-ion (lithium-ion) alternative. The price differential is considerable. The average lead-acid e-bike in China costs $167 whereas an equivalent Li-ion bike in Western Europe costs $1,546. Since so many of these bicycles' parts are imported – to a large extent from China – the Covid epidemic and ensuing hiatus in transport caused serious shortages and many bicycles are still temporarily out of commission, awaiting some crucial component. In any case all those Li-ion batteries represent part of the ever-increasing demand for the world's scant resources of rare earths as well as of lithium itself.

The same Li-ion batteries are needed for e-scooters and e-skateboards: both forms of transport rapidly increasing in popularity among the young. Because such things produce no exhaust it is easy to believe they are clean and Green – and in that sense they are. But as when considering motors running on ordinary petroleum, it is vital to factor in all the environmental costs of mining and refining the necessary components and add those to the equations surrounding the manufacture of Li-ion batteries, whether big ones for Formula E and hybrid cars, or small ones for powering e-scooters, mobile phones and hearing aids. Lord Acton's famous 'All power corrupts' referred to the temporal power wielded by prime ministers and despots alike; but 'All power costs' would be an excellent maxim for Greens to remember.

In addition to that, many people living in towns and wishing to go for walks or bike rides in the country have to drive to get there. Even if they go in electric vehicles their brakes and tyres will still add their tithe of pollution [see pages 97–100 of the Cars & Aircraft chapter]. It begins to look as though the devout environmentalist ought to remain at home and attend all sports vicariously via radio,

TV or the Internet. If so, it is a wonderful irony that the health of the planet should entail its human inhabitants adopting an unhealthy lifestyle. The truth is that the natural world would get on much better without us, as in time it undoubtedly will.

Cars & Aircraft: Hybrid, Electric and Hydrogen-powered

Ah! that pipe-dream of emissionless travel! Outside science fiction's fantasy of teleportation there is of course no such thing. However, as the implications of climate change sink in the dream has become still more potent. But before taking a look at the current possibilities for reduced emission travel there are two things worth remembering. The first is that calculating the emissions of travel must always include how the fuel is obtained as well as how efficiently it is used. The second is the Jevons Paradox.

The Jevons Paradox is one of the bugbears of environmentalism. It was named for the English economist William Stanley Jevons who in 1865 published a book, *The Coal Question*. In it he noted how James Watt's new steam engine, although technologically more efficient than Thomas Newcomen's existing engine, actually caused the consumption of coal to soar. Jevons's paradox states that if technological progress increases a machine's efficiency so that it uses less fuel to perform the same work, it will lead to increasing demand for that type of machine. Thus a larger number of more efficient machines can lead to greater overall consumption of fuel. So from an environmental point of view, and with delicious irony, a smaller number of inefficient vehicles (for example) might be preferable to a much larger number of efficient ones.

HYBRIDS

With that firmly in mind, we can view hybrid cars as an obviously sensible intermediate step between the well-understood carbon- and NOx-emitting internal combustion engine and the equally well-understood electric motor that depends on battery or fuel cell technology. They and/or purely electric vehicles are now widely seen as the practical way forward for all road transport and even aviation, with the additional (if wholly imaginary) advantage of maybe helping to break Big Oil's stranglehold. Haters of Big Oil should check Appendix 1 on page 263 for a partial list of everyday products that are ultimately sourced from oil. They will find virtually all the ingredients of modern civilised life are dependent on it and will remain so indefinitely.

In theory, this switch to electricity should lead to a marked reduction in CO_2 and other NOxious emissions. In practice, this hope is naïve if it overlooks that the electricity to power these vehicles still needs to be generated somehow. Whether it is supplied by Green but weather-dependent solar or wind farms, or else comes from power stations more reliably driven by oil, natural gas or nuclear fission – all these sources will at some point have depended on oil, either in their operation, construction or installation, and all have environmental consequences of their own. For instance, huge wind turbine blades with a life of twenty to twenty-five years are now turning out to be unrecyclable and at the end of their lives having to be cut up and buried in landfill. Certainly – and with the possible exception (depending on their design) of solar arrays – all will also have required the extensive use of concrete, which itself entails a massive global carbon burden. We are bound to add that once the plant itself has been built, nuclear fission is as clean a method as wind or solar energy of generating electricity besides being vastly more reliable, but that's another matter. The fact is that no power comes with entirely clean hands, and nor can it ever. The point at the moment is not to worry about the unattainable ideal but to concentrate on getting power as clean as we reasonably can.

SUPPLYING THE GRID

Before envisioning a future of battery-powered vehicles we need to confront a major problem: that of guaranteeing the mains electricity supply needed to keep them charged. In many countries, and above all in Europe, the demand for electricity is already outstripping supply. This will only worsen as more and more of our way of life converts to becoming reliant on it. At least part of the reason for this is the increased frequency of extreme weather.

Two obvious questions to ask about electricity as a source of power are: firstly, How is it generated? and secondly, How does it reach the consumer? Leaving aside the first for the moment, there is the problem of the grid itself. This relies extensively on powerlines that anyway entail losses due to the nature of transmitting electricity over long distances between power station and consumer. Supply is also vulnerable to increasingly violent weather. On 27 January 2021 Europe's interconnected electricity grid came close to a massive blackout due to a sudden surge in demand brought about by arctic weather. In order to save it the grid had temporarily to be split into two. Across Europe 200,000 households were left without power and much of the supply to French and Italian industry was cut. Nor is it any different in the United States. Californian utilities have long resorted to blacking-out their customers whenever especially high winds threaten to bring down powerlines and cause more of the wildfires that plague that increasingly dry state. In February 2021 three massive winter storms took out part of the grid in Texas leaving 4.5 million people without electricity and heating amid subzero temperatures. Many people had no access to water and food during the crisis and at least 200 Americans died – by some accounts as many as 700. And this in arguably the world's most technologically advanced country.

What applies in the United States applies equally to Continental Europe and the UK. The system itself is highly vulnerable. According to the head of one of the major European utilities it is not a question of if a blackout happens in some areas but simply a matter of when. Because of the nature of the grid's interconnectedness a blackout can happen even in countries with a normally reliable electricity supply. This is bad enough if a power cut catches one at home. It would be very much worse to be trapped at a motorway service station in an all-electric vehicle needing a recharge and finding a sudden major power cut has left you stranded for hours or even days. At least petrol shortages usually come with advance warning and it is always possible to carry a spare get-you-home can in the boot. You can't do that with electricity. It is conceivable that service stations might be required to install large backup batteries or else emergency generators, even if those would have to run on one of Big Oil's products.

Why is our electricity supply so vulnerable that it can be cut without warning from one moment to the next? The chief reason is that in order to operate smoothly, most European countries' generators and grids are designed to function at a frequency of 50 Hertz (60 Hz in the United States) and even small deviations from this can damage equipment. Electricity utilities everywhere constantly monitor the grid and if readings wander down to 49.9 Hz or up to 50.1 Hz technicians begin to get anxious. At 50.1 Hz they will already have begun shedding the load to protect the system and somewhere there will be a power cut. Back in January 2021, had the frequency swings not been reduced within minutes they would have caused massive and expensive damage across the entire European network, potentially dooming millions to week-long blackouts.

This may all sound a bit technical but it is absolutely relevant to why, the more electrified (and hence Greener) we try to make our way of life, the more that life will depend on hardcore technical matters involving the power supply we increasingly rely on. Ideally, we want to switch from conventional ways of generating power to sustainable if less reliable Green methods such as wind and solar, or even by burning 'renewables' such as biomass in power stations. But regardless of how the power is generated, preserving the integrity of the grid is paramount, so for the time being we increasingly build huge (and hugely expensive) megabattery installations to act as backup storage we can tap into in case of shortfall, while relying on nuclear or else conventional generators burning fossil fuel to do the donkey-work and keep the lights on.

On 11 February 2022 a trial was announced whereby the UK's National Grid could use the batteries of certain parked and recharging electric/hybrid vehicles in order to balance any deficit. If the trial is successful the scheme could be expanded so that millions of the nation's electric vehicles parked at home would act as a giant backup battery at a time when the transition to renewable (but less reliable) energy sources is complete. It is a vivid reminder of just how flaky the national energy supply may be.

HYDROGEN

Leaving batteries aside for the moment we should take a look at hydrogen because for some time now it has been the subject of increasing publicity. Hydrogen has been hyped as the cleanest imaginable fuel: an invisible atmospheric gas we all breathe small amounts of every day, the lightest element of all, Greener than Green. Combining hydrogen with an oxidising agent – possibly oxygen itself – in a fuel cell can produce electricity capable of whizzing all manner of vehicles and aircraft around leaving only a trail of hot water. This has already been achieved in limited trials, as we shall see. Yet turning hydrogen into the standard fuel for all our vehicles is unlikely to happen in any foreseeable future. There are several excellent economic reasons for this as well as those of physics and chemistry, and this is also the conclusion of practically all Western car manufacturers who instead are concentrating their present efforts on battery power. 'You won't see any hydrogen usage in cars,' Volkswagen's CEO, Herbert Diess, was quoted as saying confidently in March 2021. On the other hand some Asian car makers are actively researching hydrogen and have produced vehicles that do run on fuel cells: among them the Toyota Mirai and Hyundai ix35 Fuel Cell. In southern California (where else?) there are even some dedicated hydrogen gas stations for refuelling such cars and in Europe Germany has plans for 300 such stations by 2030.

In theory hydrogen is a great fuel despite being extremely flammable. It has the highest energy density per mass of any other: three times more than conventional jet fuel and more than a hundred times that of a Li-ion battery. Its drawback is that it has practically no mass, so has very much lower energy by volume. In order for it to be used as a gas in a fuel cell to produce electricity (or burned in a conventional engine) at normal atmospheric pressure and everyday temperatures, hydrogen must be concentrated into a liquid which then has to be kept at an extremely low temperature or under immense pressure. This pressure will probably be standardised at about 700 bars (10,000 psi). For comparison: one bar is roughly atmospheric pressure; your average car's tyre pressure is a little over two bars (30 psi), and that of liquefied natural gas (LNG) in cylinders is normally around seven bars (100 psi). So any pressure above 100 bars demands exceptional safety technology and storage in cylinders. The Hyundai's two 'tanks' take 144 litres of liquid hydrogen at 700 bars and also somewhat compromise the car's luggage space. The electricity generated goes to a 24 kWh Li-ion battery and the vehicle as a whole is over 100 kg heavier than its conventionally powered version, which reduces performance. At around £53,000 it is also very expensive. Current research into using fuel cells in vehicles to generate electricity assumes a minimum range of 300 miles in order to be reasonably competitive with battery-powered cars and this the Hyundai easily achieves. As will be appreciated, driving around safely with crash-proof tanks of potentially highly explosive liquid hydrogen under enormous pressure presents considerable technical problems. An even thornier and hugely expensive difficulty is providing a distribution system to make the supercooled fuel readily available throughout the country. This is where electric vehicles have the great advantage of being rechargeable at home. However, a significant drawback to considering hydrogen as the vehicle fuel of the future is that producing it in quantity is not Green.

There are two ways of producing pure hydrogen gas (H_2) for industrial use. At the moment virtually all of it is made by Big Oil from methane derived from fossil fuels: natural gas, coal or oil. Some 70 million tons of hydrogen are produced annually for petroleum refining and for making ammonia, most of which goes into agricultural fertilisers. An additional 45 million tons of hydrogen are used each year to make steel and methanol (the alcohol increasingly used as a component of petrol – see Appendix 2, Biofuels). In the industrial process four kilos of LPG (liquid petroleum gas), with the help of a little electrical energy, will yield one kilo of hydrogen. This method generates large amounts of waste carbon.

The alternative, 'clean' way of making hydrogen is by the electrolysis of water, yielding so-called Green hydrogen. This method uses renewable energy to separate the H_2 from H_2O. The method requires about 400 times the amount of electricity as the fossil fuel method, and at the moment barely 2 per cent of all hydrogen is made in this uneconomic way. It may sound Greener but it isn't, not least because the process requires large amounts of very pure distilled water. In theory, to get one ton of hydrogen by electrolysis you need around nine tons of water. However, not only is energy needed to pump the water to the electrolysis plant but still more is required to purify it by distillation. Generally speaking, it takes two tons of ordinary water to yield one ton of distilled water, so with all losses taken into account it actually takes nearer twenty tons of water and a good deal of electricity to yield a single ton of hydrogen by using electrolysis. In order for it to be properly Green the required electricity would have to be generated either by renewables or by nuclear power. In default, the electricity will have to come from plants that run on fossil fuels but are fully equipped with a CCUS (carbon capture, use and storage) facility.

At the moment, then, we can see that the technology for Green vehicles comes down to a choice between batteries and fuel cells. There is a third way, however. Hydrogen gas can be burned in an internal combustion engine just like petrol or diesel. And in this form it can equally be used to power a jet engine.

'GREEN' AIRCRAFT

Nearly thirty-five years ago the Russian aircraft company Tupolev modified their workhorse tri-jet airliner, the 180-seat Tu-154, and in April 1988 a Tu-155 became the world's first hydrogen-powered jet aircraft with tanks of liquid hydrogen in the tail. Unfortunately the experiment was then almost immediately suspended following the collapse of the Soviet Union, but the aircraft itself still exists and is proof that hydrogen can in practice be used to power a large aircraft. Whether it can be done both economically and safely is an altogether different question and something that only a good deal of further research will establish.

That research is currently under way in both civil and military aviation sectors. In 2019 10 per cent of the entire world's oil production went to make jet fuel, and in the same year the International Air Transport Association (IATA) announced that civil aviation worldwide had emitted some 915 million tonnes of CO_2 – just over 2 per cent of all manmade emissions in that year. It is no surprise that great efforts are being made to tackle oil-dependent aviation. Even military aviation is under pressure to change and Britain's Royal Air Force is to become carbon neutral by 2040. The UK's Aerospace Technology Institute's FlyZero project is hoping to realise zero-carbon emission commercial flight by 2030. The initial consensus on how to achieve this seems to be leaning heavily towards the Tupolev solution of using liquid hydrogen (LH_2) technology rather than batteries, but with the proviso of enormous amounts of research needing to be done and with little chance of much changing radically in the hydrogen-powered aviation scene over the next decade. The technical challenges (and investment) remain prodigious.

Electrically powered commercial aircraft are at first likely to be small and short-haul despite the aviation industry's current boosterism. Eviation's 'Alice', a battery-powered twin-engine propeller-driven electric aircraft, first flew in October 2022. It is supposed to be able to carry a payload of 2000 pounds (say nine US passengers or twelve ordinary people) for a few hundred miles. To judge from Twitter footage it was far from quiet, revealing that much of the noise of propeller-driven aircaft (including helicopters) comes not from the engine but from the blades whacking the air. In any case with a good deal of luck and better batteries an all-electric London to Edinburgh ten-seater service by the end of the decade is imaginable, although a fogged-in destination might become alarming. However, unless a revolutionary way of storing electricity is found that is far more efficient than today's existing technology, battery-powered aircraft for anything other than the shortest routes seem destined to remain a fantasy. One advantage of today's kerosene-fuelled jet airliners is that they become progressively lighter and hence more economical the longer the flight lasts. By contrast, a battery that weighs ten tons at take-off will still weigh ten tons on landing – which incidentally means designing much heavier landing gear for battery-powered aircraft, adding still further to their all-up weight and thereby limiting the number of passengers. This will not be true in the case of a propeller-driven aircraft with electric motors powered by fuel cells, of course. As the flight progresses, the tanks of liquid hydrogen and oxidiser will steadily deplete like those of conventional kerosene.

The UK's Climate Change Committee has predicted that by 2050 Britain's aviation sector will become its second-largest contributor to greenhouse gas emissions unless more sustainable methods of air travel are found. At the moment, though, much of the sector's attention is focused on developing sustainable aviation fuels (SAFs). In October 2021 Rolls-Royce and Boeing announced a partnership in which they will use a Boeing 747 'Jumbo' as a flying testbed using

Rolls-Royce's Trent 1000 engine running on 100 per cent SAF.[1] British Airways recently signed a multiyear deal to take commercial quantities of SAF produced in Britain. Initially it will be for 73 million gallons of fuel made from recycled cooking oil produced at the Phillips 66 plant on Humberside, and BA has planned to begin flights using the fuel by the end of 2022. IAG, BA's parent group, has announced a goal of 10 per cent of its flights being powered by SAF by 2030.[2] (Again, see Appendix 2, Biofuels for further information.) Maybe fatbergs will one day fly us about the world.

Several aircraft companies currently have a variety of fuel cell designs on the stocks – Airbus, for example, has the declared intention of developing the world's first zero-emission commercial aircraft by 2035, the ZEROe, that will depend in one form or another on hydrogen-powered fuel cells. Yet electric aircraft to replace today's long-haul aircraft seating many hundreds of passengers on eight- or twelve-hour flights and more may have to remain a fantasy unless the economics, technology and safety of the hydrogen component of fuel cell propulsion can be guaranteed.

What is more, the atmospheric damage caused by most long-haul flights goes far beyond the carbon dioxide emitted by burning conventional jet fuel. It is now increasingly understood how the contrails that form from the soot, water vapour and other aerosols that aircraft engines emit when flying at the most economic altitudes play a large part in heating the atmosphere and hence in global warming. Concentrating entirely on CO_2 when aiming for 'net-zero' aviation is wilfully to ignore a major part of the problem.

Using fuel cells in cars to generate electricity is quite an efficient source of power, its great advantage being that the vehicle needs no big heavy battery. Apart from the Toyota and Hyundai cars already mentioned, Honda also produced their own Honda Clarity in 2008 that was powered by fuel cell to show that the idea is practicable. However, a fuel cell will only generate electricity for as long as its supplies of hydrogen and oxidiser can keep the chemical reaction going. If either gas runs out the vehicle will silently coast to a halt. This technology, too, can go no further as a potential source of power without plentiful supplies of both gases where cars can refuel. That means dedicated outlets like petrol stations or supermarkets. It is hard to imagine kerbside hydrogen refilling points in the manner of electric chargers because the gas offers yet another problem. As the lightest element it has the ability to gradually seep through most ordinary materials, including even some metals. It is colourless, odourless and nontoxic, but highly flammable. It may not be easy to guarantee its safe use in cars and public places where people smoke and static electricity could cause sparks. Casual habits as well as technical challenges will have to be overcome before cars, lorries or aircraft can safely carry (and refuel) large tanks of liquid gas under enormous pressure, particularly if the gas is as notoriously flammable and leak-prone as hydrogen. In any case the press will be sure to dig out horrendous images of the Hindenburg disaster from its archives. This was the huge German airship that had just flown the Atlantic in May 1937 and was approaching its mooring mast in New Jersey, United States, when it suddenly caught fire and exploded in a fireball that killed thirty-five out of the ninety-seven aboard. The tragedy was broadcast live as well as filmed, and the images effectively put an end to the idea of airships for passenger travel. The safe alternative for a lifting gas is helium, which is inert; but it is also vastly more expensive than hydrogen.

For the moment fuel cells are most practicable when stationary. As noted earlier, on sites where there is plenty of space for the storage tanks of gases they need they are already used as backup for generating emergency electricity supplies if the stability of the grid is threatened.

TYRES

There is a further outstanding environmental problem associated with road vehicles regardless of how they are propelled: they all give off particulate materials from the steady wear of tyres, brakes and road surface. Tyre wear produces nearly 2,000 times more atmospheric pollution than do car exhausts. Tyres are mostly made from synthetic rubber deriving from crude oil and containing hundreds of different chemicals that vary according to manufacturer. The majority of tyre particulates are extremely small and thus are easily inhaled. Inevitably, they contain a wide range of toxic organic compounds, many of them known carcinogens. The larger particles accumulate visibly as the black dust by roadsides. Up until now, practically all monitoring of motor vehicles' emissions has concentrated on their exhausts as the most obvious visible and smellable source of pollution. Improved fuel and engine technologies, together with catalytic converters, are successfully reducing emissions of carbon monoxide (CO) and various nitrogen oxides (NOx). Glaringly unregulated, however, is pollution from other sources, most notably from tyres.

The German and Austrian equivalents of Britain's Automobile Association have for some years been testing different makes and sizes of tyres to see how much particulate matter they give off from contact with the road. They have established that abrasion from the average tyre is about 120 grams per 1,000 kilometres, which means that over its lifetime a tyre could lose up to a kilo of its original weight. An EU study claimed that some half a million tonnes of rubber and plastic are abraded from the world's tyres annually. This may be a serious underestimate since a more recent study indicates that in the United States and the UK alone some 300,000 tons of tyre rubber are shed annually as pollution, merely by cars and vans.[3] Heavy trucks and lorries are not included in this estimate and must account for a great deal more. So where does it all end up? A Dutch study in 2017 estimated that 10 per cent of all microplastic particles in the world's oceans comes from tyres. A good deal of the rest is probably in our lungs.

On 11 July 2019 the UK's DEFRA issued a press release predicting that in the next year 10 per cent of Britain's emissions of particulate matter will be from tyres, brakes and road wear. 'Particulate matter' here means particles of 2.5 microns or smaller: in other words, capable of being inhaled. Particles of this size have for some time been implicated in asthma, lung cancer and premature death. It is long overdue that this has become yet another priority. The tyre wear tests jointly conducted by Germany's ADAC – Europe's largest motoring association – and Austria's ÖAMTC have discovered that tyres vary a good deal according to manufacturer. For every size of tyre there are those with the best safety characteristics that also show less wear. Heading their lists in this respect are Michelin's, which in all tyre sizes show the least wear at 90 grams per 1,000 kilometres. At the other end of the scale are Pirelli, Bridgestone and Continental. The worst of all was found to be Bridgestone's Blizzak LM005 tyre for vans, which lost almost twice as much: 171 grams per 1,000 kilometres.[4] In general, the heavier the vehicle and the faster it is driven, the more tyre pollution it creates. All electric vehicles tend to be heavier because of their batteries. Equally, the trend towards wider 'sport'-type tyres on high-performance cars and SUVs leads all of them to higher wear and hence to more particles emitted. In the absence of legislation about the composition and construction of tyres there is little to be done beyond testing them regularly for pressure and balance to keep abrasion dust to a minimum. The particles deriving from brakes, clutch plates and tarmac road surface are another matter and constitute yet more virgin territory for activism. The new plastic-spoked 'airless' tyres are unlikely to make much difference. Chemically laden synthetic rubber will still be in contact with tarmac, each wearing the other away.

F1 racegoers might also reflect on the notoriously rapid wear of racing tyres. Some of those kilos of rubber and hi-tech composites scrubbed off all those sets of tyres in a two-hour race will inevitably drift over spectators, not to mention drivers and pit teams. And then, of course, anybody transiting through an airport is likely to inhale not only some of the gases of jet efflux but particulate matter from the heavy tyre wear (and smoke) from aircraft when they touch down. Performing or watching cars drifting or doing 'donuts' might well prove fatal years later.

Still, the original vision is unquestionably attractive of hydrogen-powered cars zipping quietly about our roads producing no emission more offensive than pure water. We are assuming the problems with the polluting dust from oil-sourced tyres, brakes and road surface will miraculously have been solved – which of course they won't. The 'hydrogen vision' also overlooks the challenge and expense of building a national infrastructure to distribute this dangerously explosive fuel. Quite apart from the safety factor, the scientific journal *Nature Climate Change* calculates that producing an e-fuel like hydrogen for cars may require up to five times more electricity than is needed for battery-powered vehicles. *New Scientist* calculated that such a switch would require about 2,000 more wind turbines[5] and the following week the same magazine compared the efficiency of battery-powered vehicles (55–60 per cent) with that of hydrogen fuel-cell power (around 30 per cent). All in all, it seems that radical e-fuels are unlikely to become cheap and abundant enough for the quick transformation that eco-warriors dream of.[6]

BATTERIES

Decades ago, magazines and comics would sometimes leap ahead to speculate about what road travel in the future might look like. Some depicted spaceship-like cars with fins driving themselves along six-lane superhighways while the family played cards in the back. Sometimes the cars were shown picking up electric power directly from strips buried in the road, a bit like the third-rail system still used by the London Underground and the New York Subway. It was all part of the pipe-dream of carefree, emissionless travel. The reality is that for the foreseeable future electric vehicles will depend on battery technology. The quest is for technological breakthroughs that will steadily make these lighter and smaller while of greater capacity and quicker to recharge.

Many people don't realise that electric vehicles have been around for well over a century. The first road car ever to exceed 62 mph (100 kph) was electric: a torpedo-shaped Belgian vehicle strikingly named La Jamais Contente that in 1899 broke the world's speed record at 65.8 mph. We forget that by the turn of the twentieth century half of all cars were battery-powered and not petrol-engined at all, that in New York alone there were 16,000 recharging stations for their lead-acid batteries. Electric runabouts like the Baker Electric of 1910 had a claimed range of a hundred miles between charging. (Those interested should check the website of Jay Leno's Garage for an episode including the Baker Electric.) Such cars were quieter, mechanically simpler, faster and easier to drive than the early petrol-driven cars. They also broke down much less and of course had no polluting exhaust. The sole reason why petrol eventually overtook electricity was because it became very much cheaper. Yet electric vehicles have survived to the present day, and in British suburban streets the quiet whine of a lead-acid battery-powered milk float may still be part of the dawn chorus.

As always, it is economics as much as technology that dictates change. Converting big commercial vehicles like forty-four-tonne lorries to electric power would, with today's technology, require batteries whose size and weight would considerably restrict the freight they were able to carry. This would push the cost of transport much higher than if they went on using conventional engines running on biofuel. However, if the price of renewable energy were to drop at the same time as the penalty for emitting CO_2 went up, then electric power for big trucks could become steadily more feasible, especially if the technology and efficiency of batteries were also to improve. As it is, electric power can already be economic for lighter trucks doing city deliveries where they can recharge overnight, and the same goes for public transport systems such as buses. But for long-haul trucking – say from Copenhagen to Naples – then barring some amazing breakthrough in battery technology to bring it closer to the efficiency of modern internal combustion engines, probably the only economically feasible way to go will be by ditching batteries altogether and using fuel cells running on hydrogen.

BATTERY RECYCLING

At present the electric cars and hybrids being produced in ever-greater numbers depend on big batteries that have limited lives, and recycling those will be energy-intensive as well as creating waste that needs safe disposal. The EU hopes that by 2030 there will be 30 million electric vehicles in Europe. The high-voltage batteries of these electric vehicles are very different from traditional 12-volt lead-acid batteries. It may seem strange to find both types of battery in today's electric vehicles: the ultramodern cheek by jowl with the positively antique. However, there is a good reason for it. There are still many conventional 12-volt accessories common to all cars such as electric windows, windscreen wipers, lighting, alarms, radio and sound systems, various computers and airbags. Having to redesign all these for a much higher voltage would make for endless complications and expense (just imagine your right-hand rear indicator bulb needing to be replaced by one of 400 volts). Those old-fashioned batteries are not only well understood everywhere in the world but are basically uncomplicated in that they consist simply of a plastic outer case, several kilos of lead, a bit of copper and the sulphuric acid electrolyte. They have long been widely recycled, although in less developed countries with cheap labour often very unsafely both for the workers who dismantle them and for the environment. The biggest danger being from lead, whether dissolved in the acid (which often gets poured away into the ground) or as fumes from melting the recovered metal and casting it into ingots for reuse. More sophisticated mechanised techniques simply shred the batteries and reduce the dry parts to powder from which the lead is then extracted, either by high temperature melting or chemically by dissolution.

At present, almost none of the newer Li-ion batteries in electric cars and hybrids are fully recycled anywhere. Larger than lead-acid batteries, they are also much more complicated, being made up of many hundred individual Li-ion cells, each of which needs to be individually dismantled. These contain hazardous materials and if wrongly taken apart can even explode and cause a fire that proves extremely hard to put out. (In April 2021 a Tesla 'S' model crashed and caught fire in the US, burning to death its two occupants, neither of whom had been driving. The fire took four hours and 30,000 gallons of water to extinguish.) Apart from cars and buses, there are of course rapidly growing sectors of transport and leisure that also use Li-ion batteries: in e-bikes, e-scooters and a wide range of electronic equipment such as power tools, down to the button batteries in bathroom scales and hearing aids.

The great irony here is that despite European car manufacturers producing electric vehicles such as the Renault Zoe or the BMW i3, they are almost completely dependent on Asian companies for the supply and manufacture of the materials for their batteries, as mostly for the completed batteries themselves. Asia produces 90 per cent of the world's car batteries, at least half of which come from China. Ideally, recycling these newer-generation electric vehicle batteries will recover the cathode metals such as cobalt, nickel, lithium and manganese. (Any aluminium and copper in the casing goes into normal recycling.) At the moment, probably only 5 per cent of all Li-ion batteries are recycled, and crudely at that. In many countries the percentage is zero and the batteries are simply dumped. Because each one of the hundreds of cells needs to be taken apart, dismantling them is very time-consuming. When it is done at all, most of the battery is simply reduced to what the trade refers to as 'black mass': a mess of lithium, manganese, cobalt and nickel that then needs to be processed to recover each element in usable form. This itself is energy-intensive. Painstakingly dismantling the fuel cells by hand does indeed make it possible to recover more of each element but in addition it brings other problems, notably the cost of man-hours. This may not present a serious problem in countries where labour is comparatively cheap and where environmental contamination and working conditions are permitted that would not be allowed in Europe or the US. In short, the economics of recycling Li-ion batteries means that the process cries out for automation. Robots are definitely the way to go to recover the maximum amount of the valuable elements locked up in these electric vehicle batteries. The UK has no indigenous supplies of any of these raw materials in industrial quantities, although in Cornwall lithium is increasingly being extracted from geothermal fluids and also from the mica associated with that county's abundant granite. Copper, too, has recently been discovered in exploitable quantities.

For the moment, the lithium on which the Green movement has to pin at least some of its hopes comes from two main sources: either from the mined ore spodumene – of which it takes 250 tons to make one ton of lithium – or from suitably mineral-rich brine. This last comes typically from beneath Chile's Atacama salt flats. It takes 750 tons of brine to make a single ton of lithium, which also requires 1,900 tons of pure water to extract. Sixty-five per cent of all the region's water goes to mining, leaving local farmers having to bring in what they need from elsewhere.[7] Another essential element in electric vehicle batteries is nickel, much of which comes from the world's largest reserves in Indonesia. As usual, such mining is intensively pursued with full government support and scant regard for the health of either the environment or the local people. An investigation by the *Guardian* into nickel mining and the electric vehicle industry has found evidence that a source of drinking water close to one of Indonesia's largest nickel mines is contaminated with unsafe levels of hexavalent chromium (chromium 6). This is the cancer-causing chemical more widely known for its role in the true-life Erin Brockovich story and film, in which an environmental crusader exposes a power company that polluted the tap water in Hinkley, California with chromium 6. The *Guardian* also found evidence suggesting elevated levels of lung infections among Indonesians living close to the mine.

At the moment the economics of these new vehicle batteries are very hard to pin down, partly because there are still comparatively few electric vehicles on the roads and it is not yet a mature market. Another reason is that battery technology is in a state of almost frenzied development with new designs and chemicals suddenly becoming topical. A recent design by the Chinese battery company BYD looks like a breakthrough in that it seems superior in energy density to offerings by Audi, Jaguar and Tesla. But who knows? Nor does it help market stability that the price of rare earths and other raw materials often fluctuates wildly. This in turn can make recycling a dubious investment. To take one example, the price of cobalt tripled between 2017 and 2019 then dropped abruptly before rising once more, although not to its previous level. There is now a growing use of nickel, which may well gradually take the place of cobalt, at least in the short term. From a potential investor's point of view it all makes for a volatile and unsatisfactory market. And of course this is conveniently to pretend one doesn't know (or care) about poisoned Indonesians and the abominable child slavery in Chinese-owned African cobalt mines.[8]

INFRASTRUCTURE

In their optimism, many Green consumers seem oddly unaware of the basic law of physics that presumably holds good everywhere in the universe: You cannot get something for nothing. In theory it would be perfectly feasible to run a car on water; but turning H_2O into a usable fuel requires a large input of energy from another source. Even as things are with hybrid cars and electric vehicles, recharging them requires energy to be generated somewhere and made readily available everywhere: a massive investment in infrastructure. In the UK, charging points will need to be ubiquitous well before the government's ban on new petrol and diesel vehicles comes into force in 2030. At the moment, to nobody's surprise, the provision of electric vehicle charging points is heavily skewed in favour of London. In 2021 London and the South-East received 45 per cent of the entire country's new chargers. The Society of Motor Manufacturers and Traders has already estimated that Britain will need to build the best part of 2 million charging points by 2030 – a rate of well over 500 a day at a cost of £16.7 billion. Nobody familiar with modern Britain would care to bet on that ambition being fulfilled. Nor would they be surprised that at the moment fourteen different private companies own charging networks, headed by bp pulse, but with no single company having more than 13 per cent of the market share. The immediate result is grotesque inefficiency and inconvenience. As the journalist Ros Coward reported after a year of driving an electric vehicle:

> Each provider requires you to download its app and subscribe. They all work differently and prices vary. Many chargers are out of order. As often with electric vehicles, there is a catch. Some borough councils do not reserve the parking spaces for charging electric vehicles, so 'ordinary' cars park there, displacing electric vehicles, which are then forced to hunt around for one of the faster, more expensive chargers [...] Conscious of conserving energy, you drive at lower speeds, while travel by train for longer journeys becomes more appealing.

But the unnecessarily complicated charger network leaves everything to be desired. There is no universal charge card for all suppliers and far too few chargers overall. This will be a huge problem as electric vehicle numbers increase. Lack of government regulation is also ridiculous: as things stand, multiple suppliers face no comeback if they fail to keep chargers working.[9]

It is all so predictable. The goal of getting vehicles to go electric is undoubtedly right and worthy. Yet instead of the Department for Transport and the government laying down in an organised way what is arguably the most vital piece of Britain's infrastructure since 1945, the recharging scene is already a chaotic free-for-all, with privatised suppliers fighting for market share while most of Britain is predictably left behind the South-East of England. If it is any consolation to Britons, Germany and Austria have also opted for a similar competitive model of charging points that are not interchangeable and are furnished by privatised suppliers.

As so often with essential public services this is the wrong economic model for efficiency, with the wrong politicians in charge. Is this too cynical? Absolutely not. In fact it's nothing like cynical enough, especially when one comes to compare this new scheme's chances with efforts to build an infrastructure to provide the whole of Britain (and not just the London area) with reliable mobile phone coverage. It is both germane and instructive to look briefly at how that has gone so we can make some predictions about the provision of charging points for the country's electric vehicles in the next few years, while at the same time always bearing in mind the potential frailty of the power supply to national grids mentioned earlier in this chapter.

In the UK in December 2021 the severe storm Arwen left some communities without electricity for nearly two weeks. Most of the worst affected were in northern Britain where in many places mobile phone reception is still so poor that even in normal times locals need to drive or walk to the top of a nearby hill in order to get a signal. These are often areas where landline telephones have been brutally 'phased out' (meaning British Telecom has simply discontinued the service, forcing people to rely on mobile phones). The BBC reported one lady in the Lake District as saying: 'The weak phone signal I struggle with at the best of times was knocked out by the storm, as was the wi-fi I use to boost it. There was absolutely no signal in the whole of the valley for three or four days, and I couldn't drive up the hill to get one because the roads were all closed.' Regular power cuts make it difficult for her to live and run a business because the communications system relies on electricity. 'It's embarrassing that a supposedly world-leading country has such a shonky infrastructure. I had full 4G in the mountains of Transylvania a few years ago.'

The MP for Westmorland and Lonsdale, Tim Farron, says having no phone at all is downright dangerous for lots of people. 'Putting all your eggs in one basket isn't a good idea. If the power goes out and you have no means of charging a phone then you're in trouble.' In Rothbury, Northumberland, all three mobile masts serving EE, Vodafone and O2 failed in Storm Arwen. As Steven Bridgett, the county councillor for the area, asked: 'Why don't those masts have back up generators?' He spotted the irony that the Green agenda is not taking into account the realities of weather associated with climate change. 'Electric cars, air source heat pumps and the new digital landlines are all useless in circumstances like this. The whole thing's an absolute joke in the aftermath of a storm like that.' He wants resilience built into utility infrastructure by the organisations responsible for it. 'That's what they're there for and what people pay their bills for.'[10]

Alas! The country is no more that over eighty years ago could quickly organise and impose a rationing system for all its then-55 million citizens, while at the same time building more than 450 new airfields, each one needing 50 miles of drainage ditches and 130,000 tons of concrete, in all covering some 36,000 acres.[11] That we have lost a national ability to organise and be businesslike is perhaps no surprise when a modern British prime minister, Boris Johnson, as a member of the Conservative Party that traditionally espouses business and industry, famously said 'Fuck business!' in conversation with Airbus's Head of UK Publicity. No surprise, maybe; but still a cause for national shame. We have become the Monkey that can't even find the jar with the banana in the middle of a jungle clearing. As already mentioned, the UK will phase out the sale of all new petrol- and diesel-driven cars by 2030. Current sage advice to millions of citizens would be to forget going Green and to hang on to their old cars until they fall apart.

The Fashion Industry

By United Nations' reckoning the fashion industry is the world's third biggest manufacturing sector. On its own it uses more energy than the shipping and aviation industries combined, accounting for some 10 per cent of all carbon emissions.[1] The World Economic Forum agrees, saying that fashion is the third most climate-damaging industry after food and construction and ahead of cars, electronics and freight. Western consumers are currently buying 60 per cent more clothing than they did a mere fifteen years ago. Apparently, the average British woman buys half her bodyweight in clothes annually and is likely to have twenty-two unworn garments in her wardrobe. Every year, the average Briton buys 26.7 kilos of new clothes: almost exactly twice what Italians buy.[2] In the United States the average consumer buys an article of clothing every five and a half days, or roughly sixty-six new garments each year, and it is claimed the average American woman owns thirty outfits. In any case fashion houses are now producing twice the quantity of clothes they were offering in 2000 and in the next eight or so years this is forecast to rise from today's

62 million tonnes to 102 million. In terms of weight this would be equivalent to an additional 500 billion T-shirts. Assuming the average length of a T-shirt is fifty centimetres, this would mean that if all 500 billion were laid head-to-tail they would stretch the distance between Earth and moon sixty-two times.

To someone who tends to wear the same clothes until they fall apart and quails at the thought of the changing-room antics involved with having to buy a new pair of trousers, these statistics are almost incomprehensible. It is even more astonishing that within a year most of these new clothes will wind up in landfill. More statistics: the average Australian buys twenty-seven kilos of new clothing annually and dumps twenty-three of them to landfill, accounting for 93 per cent of all the textile waste that country generates. These figures are second only to those for US consumers.[3] Nor does the UK lag behind. According to UN figures quoted by the BBC, three out of five fashion items wind up in landfill within a year of purchase. This is indeed 'fast fashion'.

The March 2021 issue of the Royal Geographical Society's magazine *Geographical* assessed the environmental footprint of the worldwide clothing industry and agreed with the UN's figures that, on its own, this sector is responsible for up to 10 per cent of global carbon emissions. Assuming for the moment that jeans and T-shirts are made entirely of cotton (which many aren't), producing the fabric involves growing the cotton, picking it and spinning it into thread, knitting the thread into fabric, dyeing it, cutting and sewing the garment (adding zips or buttons and rivets) and then transporting the finished article to the point of sale – usually halfway around the world since most production takes place in less developed countries where labour is cheapest.

COTTON

Cotton is grown in some eighty countries and is among the thirstiest of all agricultural crops. It has been estimated that to grow enough cotton to make a single T-shirt can require as much as 7,000 litres of water – in the case of a pair of jeans 7,500 litres (1,650 gallons), or enough water to fill eighteen large bathtubs to the brim. About half of this will go into the ground to irrigate the cotton plants and the other half into subsequent factory processes. Cotton is also the world's most chemically sprayed crop and accounts for 16 per cent of all insecticide releases,[4] although UN sources put that figure higher at nearly 25 per cent. By far the world's largest textile producer is China, and the majority of the dyed cotton fabric used to make the denim of the world's jeans originates there. In China the electrical power for this high-energy process is practically all generated by coal-fired power stations. The EU produces almost half as much fabric as China – still a prodigious amount. Spinning yarn and weaving it into fabric also requires various chemical additives for lubricating the fibres and eventually for waterproofing. Persistent molecules of the various chemicals used in these processes are now ubiquitous, found at both Poles and readily taken up in the body fat of seals and polar bears.

The fabric is then dyed: a process that requires not only quantities of clean water but energy to heat it because most dyes need temperatures of at least 60°C in order to become fast. In the case of jeans, the industry uses 50,000 tonnes of synthetic indigo dye as well as other chemicals. Much of the waste water from these factory processes finds its way into groundwater and rivers and hence eventually into the sea. Increasingly, and especially in the case of T-shirts, many additional dyes and inks will be required to print slogans and designs on the garments. Once dry, the fabric is sent off to the world's (largely Asian) sweatshops to be turned into clothing. 'We don't talk about child slavery here,' a Bangladeshi factory manager was quoted as saying. His interviewer sarcastically suggested the alternative phrase, 'Regrettable Economic Necessity.' 'Much better,' the man agreed. 'We are a poor country, not an inhumane one.'[5] Although the throughput to supply the global fashion industry's myriad different sizes and styles is huge enough, up to 15 per cent of the fabric will nevertheless be wasted in the form of offcuts for which until recently there has been no use, although as acknowledged later in this chapter efforts are now being made to use at least some of this wastage. Finally, the finished T-shirts are shipped around the world, mainly in container ships but increasingly also by air in order to keep pace with customers' ever-growing habit of online buying and insistence on fast delivery. The popularity of this form of shopping has burgeoned recently, at least partly as a result of the Covid lockdowns that have also hastened the demise of so many high street outlets starved of customers and unable to pay the rent on their premises. *Geographical* estimated that moving a mere 1 per cent of garments from maritime to air cargo could result in a 35 per cent increase in carbon emissions. (See the chapter on Shipping & Shopping.)

SYNTHETICS

An important part of the environmental impact of the worldwide industry making clothes (or 'apparel', as the US likes to call it in its charming old-fashioned way) comes not from cotton but from synthetic fabrics. Synthetics date from the late nineteenth century when three chemists working in Kew on the outskirts of London made the first usable rayon out of cellulose extracted from wood pulp. They patented this in 1894 and it was widely successful as artificial (or 'art') silk. The three chemists quickly sold their invention to Courtaulds who from 1905 became the world's largest producer. Art silk soon acquired alternative names such as viscose and rayon although industrial chemists thought of them as polymers. These went on being produced for the next ninety years or so while other polymers were invented using petroleum-based chemistry. In the late 1930s the US company DuPont invented a polymer they patented as 'nylon', which counts as the first polyester. The next was synthesised by a group of British chemists in 1941 who patented theirs as

Terylene. In 1946 DuPont bought up all rights to Terylene and invented another, slightly different, polyester they called Dacron. Since 1950, petroleum-based fibre production has enormously increased and many different variants have been produced, up to and including modern 'luxury' polyesters like microfibre as well as 'sporty' ones like neoprene. Natural fibres like cotton and flax remain more expensive than synthetics because labour and production costs are so much higher even without the stiff competition for scarce land on which to grow them. Polyesters are much simpler to produce, three of the four main kinds being made from oil – about 70 million barrels' worth each year. An increasingly important 'stock' for making polyester comes from recycled plastic bottles that themselves originate from oil. The fourth type of polyester can be made from organic sources such as wood pulp. Ultimately, all synthetics depend on some very artful industrial chemistry.

Back in the 1950s woven polyester fibres produced a material that was widely hailed as 'labour saving' because it went on looking good when worn for several days without being washed and ironed. For a while 100 per cent nylon shirts became fashionable for businessmen and were much touted as providing the crisp, Jet Age look for thrusting entrepreneurs. With them the phrase 'drip dry' entered the language. However, it was not long before their drawbacks became apparent. Not only did the material feel slightly slimy when worn, it turned out to be smelly and uncomfortable in hot weather because nylon isn't 'breathable' and promotes sweating. Also, like certain of their wearers, white nylon shirts began to take on a faintly yellowish tinge as they aged. As a consequence, pure nylon fabrics largely went out of fashion and were replaced by other, more comfortable polyesters. The advantages of a blended fabric like a cotton/polyester mix include being harder wearing, more windproof and more resistant to wear and tear. However, one serious disadvantage of polyester quickly led to its being banned from undergarments worn by military aircrew and from combat fatigues in general. It was soon discovered in the Second World War that while cotton does not burn easily but merely chars, polyester melts and welds to the skin. This makes it all the more surprising that until at least the 1960s sheets, petticoats and even children's nightdresses were often made of nylon that readily went up in flames if they touched an electric fire, leaving wearers with horrendous burns that many didn't survive.

Any polyester shirt has double the carbon footprint of one made entirely of cotton. Apart from those 70 million barrels of oil that each year go to make the majority of polyesters, an additional environmental disadvantage of these and other oil-based synthetics is that nobody yet knows for certain how long they will take to break down in the soil. In the case of a cotton/polyester mix the cotton will rot away within decades, leaving the polyester component. Some estimates predict this will need at least 200 years to decompose, others that it will take several centuries longer. As it slowly breaks down, its residues will be as toxic as those of any other oil-based plastic. Given that synthetic materials like polyester, acrylic and nylon represent around

60 per cent of all clothing materials worldwide,[6] this will have alarming consequences for the mountains of dumped clothing that increasingly litter distant shores. Even while in use they shed microscopic plastic fibres when washed and especially when tumble-dried, the majority of which eventually find their way via wind, rain and rivers into the sea. It is currently thought that synthetic fibres may account for up to 35 per cent of the microplastics in the world's oceans. Again, nobody knows their lifespan, any more than they know whether they are harmful in lungs and bodies.

It may be remembered that the earliest of the four main polyester types – viscose or rayon – was environmentally friendly in the sense that it was not sourced from oil but from wood fibre. In the early 1980s Courtaulds' chemists managed to find a way of making it that did not generate a pungent effluent that could make living downwind from a rayon factory very unpleasant. The new fibre was known generically as lyocell and was patented as Tencel. Where I live in Austria local traffic is frequently held up at a level crossing to allow a seemingly endless train of goods wagons to make its way to the only major local factory, a few kilometres down the line at Lenzing. As one sits and waits for it to pass it is saddening to watch the truckloads of hardwood tree trunks – notably beech – go past: the torsos of what until recently had been noble forest giants, mostly in Poland. This Lenzing factory has, over more than a century, grown from a wood pulp and paper mill into being a major global producer of wood-derived synthetics. These range from viscose, acrylic, polyester, and their flagship material, Tencel, which was sold off when Courtaulds was broken up in the 1990s. Lenzing has also recently developed a way of transforming some of the clothing industry's hitherto unusable cotton offcuts into a material they call REFIBRA. This involves repulping the cotton and adding it to the wood-based polyester to make Tencel: clearly a step in the right direction. Together with plans for installing acres of solar arrays, the Austrian factory that was once a major local polluter making viscose in the old way has been transformed into a consciously Green enterprise – which is hopefully true also of the group's overseas branches in Grimsby, UK and in Mobile, Alabama. Even so, although these cellulosic fibres are biodegradable, once they are in landfill they generate methane, which makes their environmental footprint only a little smaller than that of polyester.

Even so, and for all that 'sustainability' has become a word of quasi-religious significance across all manufacturing industries, the sight of those countless tons of beautiful hardwood trundling past to be turned into three-piece suits, slacks and sportswear still does not feel right. It is all too easy to imagine swathes of forest felled for the insatiable global rag trade without being wholly convinced that fresh saplings really are being planted at the same rate and allowed to reach maturity. Is this 'sustainably sourced timber'? Maybe it is. But it feels even less right to know that so much of the clothing the trees have died to make will wind up, some of it not even worn, on a tottering mountain of discards in a far country. When considering these global industrial sectors, environmentalists often tend to focus their attention far too exclusively on the manufacturers. The time has surely come to pay just as much attention to consumers and their unreflective, casual greed to acquire that is matched by an equally casual swiftness to throw away.

SHOES

Shoes are an important and lucrative element of fashion, and over the last decades have increasingly reflected the gradual encroachment of sports and leisure activities into everyday wear. Here again the industry's sheer output is matched by shoppers' spendthrift habits. Each year worldwide some 24 billion pairs of shoes are produced for the world's 8 billion people, of which the trade casually accepts that 90 per cent are likely to be thrown away within a year. Because of its cost, genuine leather had become something of a niche component of modern shoes until major sportswear brands rediscovered its advantages in terms of 'breathability', comfort and (as they believed) its credentials as a 'natural' material that is wholly biodegradable. A recent report revealed the degree to which this environmental wishful thinking can amount to Greenwashing.[7] It turned out that over fifty fashion brands of shoes had multiple supply-chain links to the largest Brazilian leather exporter, JBS, which is known to engage in Amazon deforestation to extend its cattle ranches. The brands included Coach, LVMH, Prada, H&M, Zara, Adidas, Nike, New Balance, Teva, UGG and

Fendi. Quite apart from the ongoing rape of what remains of the Amazon forest, 'natural' leather carries its own high environmental price in the form of farting, biomass-consuming beasts and the highly polluting tanning and dyeing industries. It should be noted that Brazil is also the source of much of the leather that goes into the 'bespoke' upholstery of upmarket car marques: yet another sector that deserves closer scrutiny.

Probably the most important segment of the global shoe market is for quasi-sports trainers ranging from running and jogging shoes, through basketball and football boots to what the US market calls 'sneakers'. It has all been made possible by industrial chemistry. Out have gone the thin canvas 'gym shoes' of my own youth, soled as they were with genuine rubber. Their place has long been taken by sports-style footwear: a trend that is now the norm everywhere. Today's senior citizens shuffle about in trainers like retired and broken athletes unable to let go of the past, while wealthy society ladies can drift about their Manhattan haunts in absurd golden basketball boots by Louboutin. But irrespective of fashion, these modern sports shoes are undoubtedly practical. Even at the cheaper end of the market they are light, comfortable and generally hard-wearing.

This is all well and good; but the problem is that all these kinds of footwear depend absolutely on materials derived from petrochemicals, as do the shock-absorbing soles of expensive shoes whose uppers are made of genuine leather. Shoes that must above all be light, and their soles cushioned, almost always contain foams such as ethylene vinyl acetate (EVA) and thermoplastic polyurethane (TPU), both of which have their origin in crude oil. Such foams are indeed light and shock-cushioning, but their ability to biodegrade is virtually nil. It has been estimated that a typical lightweight shoe sole might take a thousand years to decompose. What is more, making such materials is polluting and energy-intensive. Recent research by the Massachusetts Institute of Technology estimates that on average, making a single pair of running shoes produces between 11.3 and 16.7 kilos of CO_2 – which is a lot for a non-electronic product: the equivalent of driving a petrol or diesel car some fifty-two miles. Naturally the packaging, shipping, advertising and storage all add to this carbon tally. The big sportswear companies (Adidas, Nike, etc.) are at last making an effort to use recycled materials where they can, but it is not proving easy.

EVA is also familiar in the form of shower clogs, garden kneelers and the soles of flip-flops. Flip-flops are the preferred daily footwear for millions of rural people in tropical places like South-East Asia. In one of my books I described being in a small boat in Philippine waters and coming upon a rocky islet buried beneath thick deposits of flip-flop offcuts. These were sheets of EVA foam holed where the soles had been punched out in a factory, presumably somewhere fairly local. Instead of being recycled the sheets had been dumped or had otherwise found their way into the sea where local winds and currents kept piling them up on this islet. For all I know they are still there, although wave action would eventually break them up into granules that in turn would become part of the oceans' incalculable freight of plastic waste.

SPORTSWEAR

One of the biggest – and fastest growing – markets for synthetic fabrics of all kinds is that of sportswear and equipment. The whole fitness/wellness market has burgeoned in recent decades, bringing high fashion as well as new technology to sportswear of every kind and stamping its logos on every last sweatband. This has spawned new hi-tech fabrics, especially those designed for both lightness and warmth as well as to repel external moisture while wicking sweat away from the body. These are almost all synthetics. Wool is old-fashioned and cotton gets damp and chilly (not for nothing does the hiking and climbing fraternity refer to 'killer cotton'). In their place have come materials such as Spandex, Lycra and Elastane. These particular three are actually the same stuff under different countries' trade names: a nonrecyclable copolymer that was invented in 1958. They offer comfort with considerable stretchability and have become important components in a whole range of sports- and fashion-ware to the extent that in the UK bad-mannered cyclists are often referred to as 'Lycra Louts'. This is yet another ubiquitous oil-derived modern material that is not biodegradable. Even after their ageing former wearers have long forsaken their bikes for walkers, the form-hugging cycling suits that once showed the workings of their every muscle will not have broken down appreciably. They will still be perfectly identifiable in their landfill graves centuries after the lout himself has broken down into dust and gas.

Another manmade material used by both the sports and fashion industries is neoprene. Older even than nylon, this was also a DuPont invention dating back to the early 1930s. As a tough synthetic rubber this polymer has long been used to make such things as hoses, gaskets, fan belts and waterproof work boots. It is precisely because neoprene doesn't perish like natural rubber that it has found such a wide range of uses. When it was found that it could be foamed with nitrogen so as to contain tiny bubbles it became widely used where temperature insulation is important. This is why neoprene is the standard material for wetsuits, although it needs to be progressively thicker the deeper its intended use because external pressure squashes the foam, reducing its insulating properties. Millions of wetsuits are in use the world over. Yet raw neoprene need not be foamed and can also be used as a skin-hugging fashion material to make tights, not to mention a whole range of squeaky fetish garments for bondage and S&M devotees.

The normal decomposing rate of neoprene is negligible and many divers and surfers are aware that the manufacture of their wetsuits' material is the very opposite of 'Green'. They try to compensate by 'upcycling' their old suits, perhaps by making things like mouse mats or tablet sleeves out of them. In Australia, at least, there is an emergent technology for turning old neoprene into bitumen. Maybe one might rephrase that as 'back into bitumen' since both ultimately derive from petroleum, itself the natural product of millions of years' worth of organic material subjected to heat and pressure. Elsewhere, though, old wetsuits tend to wind up in rubbish dumps and a putative scavenger in 500 years' time might unearth one and find it scarcely changed.

Still worse, a good many dumped clothes include those that, like Gore-Tex, will have been sprayed with chemicals intended to make them impermeable to water. These are polyfluoroalkyl substances (PFAS). They form a large group of chemicals that has been growing since their discovery in the 1940s. They are linked by their mixing of the two elements carbon and fluorine that together form one of the strongest bonds in all organic chemistry. This is the origin of their collective name 'Forever Chemicals' because nobody knows how long it takes for PFAS to be broken down. They are ubiquitous in thousands of industrial applications that include coatings made to stop food sticking to saucepans, preventing carpets and furniture from being stained by water, and in clothing designed to be weatherproof. PFAS molecules are now being found in food and also in the human body, and much urgent research is being conducted to discover what – if any – long-term damage or susceptibility they might cause.

'NATURAL' TEXTILES

There is now some hope in the efforts currently being made to discover new sources for textiles that are less reliant on the wizardry of industrial chemists. These include such things as fabrics made from pineapple leaves, mushrooms, yeast-based proteins and yarns derived from sugarcane and other kinds of biomass such as wood pulp, as in the case of Tencel. Pineapple fabric or piña has been made in the Philippines for centuries although much of the industry lapsed in the Second World War and until recently its revival was slow. The fabric is made from the long fibres in pineapple leaves which, when woven, produce a light, filmy, sheer material much used to make the traditional barong tagalog. This is a tailless shirt, usually long-sleeved and often beautifully embroidered. It is both elegant and supremely cool in a hot climate. As may be expected, given the skill and labour involved, these are not cheap garments. Also, because the material is semi-transparent, it is particularly suited to flattering slim Asian physiques rather than the 'dad bods' of doughy Westerners. No doubt some of the other sources of natural yarn will yield denser fabrics with other advantages – such as being easily dyed or hard-wearing.

'INTELLIGENT' TEXTILES

At the polar opposite of such naturally sourced materials is the next generation of novelty fabrics often referred to as 'intelligent', possibly to distinguish them from their wearers. E-textiles are designed to respond to environmental stimuli such as heat or light by changing colour. There have long been schemes for turning clothes into computers for such things as keeping track on the wearer's physical health, although I have yet to see references to a baseball cap designed to assess mental health. Aiming a bit lower, the Nadi X yoga pants from Wearable X have accelerometers and vibrating motors woven into the fabric around the hips, knees and ankles that gently vibrate to instruct wearers on how to move, should this prove too much for their intellects. Even more futuristically, the College of Optics and Photonics at the University of Central Florida has announced the first user-controlled colour-changing fabric that enables wearers to change its colour by using their mobile phone. The College describes it thus:

Each thread woven into Chromorphous's fabric incorporates within it a thin metal micro-wire. An electric current flows through the micro-wires, slightly raising the thread's temperature. Special pigments embedded in the thread then respond to this change of temperature by changing its colour. Users can control both when the colour change happens and what pattern will appear on the fabric using an app. For example, a solid purple tote bag now has the ability to gradually add blue stripes when you press a 'stripe' button on your mobile phone or computer. This means we may own fewer clothes in the future but have more colour combinations than ever before.[8]

ACTION MAN CHIC

Fabric with wire in it sounds less than easily biodegradable even before we know what the material itself is. Maybe such things will become the height of fashion, maybe not. Of more immediate interest are the current trends for rugged clothing, which no doubt reflects the law that says the more humdrum urban life becomes, the more one should dress as for a commando raid on Tracy Island. First came 'warcore', which was essentially Action Man kitting himself out for imminent geopolitical meltdown. Typical were desert-style scarves and waistcoats that looked like flak jackets. Army surplus chic had, of course, been around since the Second World War, but it relied on garments that mostly were genuinely ex-services and made of traditional fibres like cotton, wool and hemp. It was a look that mostly came to signify student poverty (at university I had an excellent-quality Civil Defence overcoat in navy wool bought for peanuts). Today's warcore as a fashion relies far more on manmade fibres and is usually evidence of a bank card comfortably far from being maxed out.

As I write this in late 2021 a fashion known as gorpcore seems predominant. The name derives from the 'trail mix' favoured by US hikers: 'Good Ol' Raisins and Peanuts'. What started out as utilitarian and functional has acquired expensive outdoor labels for the challenging city environment and involves hi-tech winter jackets and storm-proof trousers for a look suggesting that the wearer, despite being in Soho, is just about to climb K2. Frostbite-preventing underlayers and advanced Gore-Tex in Central London probably means a recent visit to shops selling the Icelandic brand 66°North or else the Swiss Mammut. Invisible molecules of 'Forever Chemicals' swirl in the wakes of these intrepid urban adventurers. This is urban cosplay. Thirty years ago their counterparts would probably have worn waxed cotton Barbour jackets and a chequered cap to suggest a Range Rover parked around the corner, spattered with the mud of their imaginary country estate. And even further back in the 1960s cosplay was a regular entertainment in The Coleherne, a gay pub in Earl's Court that catered for the leather crowd. In they came on Saturday nights, studded black jackets creaking and swinging a motorbike helmet like Hell's Angels looking for action. After closing time some might be glimpsed standing furtively outside with their helmet, waiting for the bus back to Fulham. Human nature doesn't change and least of all when fashion is at stake.

DUMPING

It is the persistence of so many of today's synthetic materials that makes the vast tonnage of fast fashion clothes being dumped annually so significant. Each year 39,000 tons of discarded but often nearly new clothing are dumped in Chile's Atacama Desert.[9] Similarly, African countries such as Ghana are 'drowning under the weight of waste clothing shipped to their shores every week from Europe, America and China.'[10] In both cases drone photography reveals prodigious alps of variegated clothing: Mount Trashmores of sheer waste. In Ghana alone, 15 million items of often barely worn clothing arrive each week. Although some of it gets quickly picked through by traders on arrival, a good 40 per cent of it ends up on the coastal landfill mountain steadily slumping into the sea: another rich source of polyester that in fragmentary form winds up in marine life, and thence in the people who eat the fish.

The reason why Africans and Chileans are unable to benefit to any extent from all this often new second-hand clothing is because much of it consists of rubbishy fast-fashion items whose quality is too poor to make it worthwhile trying to sell on. The BBC News claimed that 'the fashion industry loses $500 billion a year due to fashion waste.' We take that to mean the retail cost of all the fast-fashion clothes dumped annually. So what has happened to bring about the astounding statistic quoted in this chapter's first paragraph: that the average Briton buys almost twenty-seven kilos of new clothes annually? Why are today's Western consumers buying 60 per cent more clothing than they did a mere fifteen years ago?

There may not be a simple answer to this although the great increase in mobile phone and tablet ownership, together with more money and burgeoning social media, might have a good deal – or nearly everything – to do with it. With social media have come the influencers with the power to start and promote a particular fashion. At the same time the fashion industry has greatly quickened its response to new trends and has both speeded up its ability to manufacture as well as to market and distribute a new line. The ease (from the consumer's point of view) of online shopping has surely helped to do the rest.

(The Monkey eyes the banana in the jar and wonders whether it would be worth waiting for a mango. Or what about a pawpaw or a bunch of rambutans? As its mouth waters the humble banana in the jar takes on a variety of colours and shapes until it becomes a general blur of desire, and in goes the Monkey's hand.)

Military Carbon

When various experts make their assessments of global environmental damage it is noteworthy that until very recently the prodigious fuel consumption and pollution by the world's armed forces was scarcely ever alluded to and even today is almost never included. The carbon deposited in the atmosphere by military aircraft alone is invariably overlooked in favour of making taxpaying citizens feel guilty about the flights they take in civil aircraft. It is more than a little absurd that we civilians worry about filling our virtuous hybrid cars with a few gallons of petrol when the world's military is burning off thousands of tons of fuel hourly in cavalier fashion – and also at taxpayers' expense.

According to Scientists for Global Responsibility (SGR), the world's militaries and the industries that supply them together account for 6 per cent of all global emissions.[1] At a time when most nations and industries are making at least some effort to cut back on emissions, many of the world's armed forces are actually increasing them by raising their spending. The Stockholm International Peace Research Institute (SIPRI) assessed the world's total military spending in 2020 as $1.9 trillion (£1.5 trillion), an increase of 2.6 per cent on the previous year. SIPRI listed the top five spenders as: the United States ($778 billion), China ($252 billion), India ($72.9 billion), Russia ($61.7 billion) and the UK ($59.2 billion). All these countries increased their military budgets in 2020.[2] It turns out the United States spends more on its military than China, Russia, India, Britain, France, Saudi Arabia and Germany combined. Given its 4,800 military bases in 160 countries this is perhaps not surprising.[3] In view of this it is equally unsurprising that in 2019 Brown University's 'Costs of War' project cited the US military as 'the largest single source of greenhouse gas emissions in the world'. By comparison with such figures the £60-odd billion that Britain spent on its armed forces in 2020 seems a paltry sum, even though it would have made all the difference to the country's failing health and education services had it been better spent.

The reason why such vast budgets can generate 6 per cent of all global emissions without being called to order is at least partly traceable to a loophole in the Paris Agreement (aka the Paris Accords) of 2015. Back in 1997 the world's militaries were granted exemption from having to meet CO_2 targets under the Kyoto Protocol, although it was never made clear exactly why. This concession was ended by the Paris Climate Accords in 2015; but two years later President Trump unilaterally withdrew the US military from this pact even though it was one signed by practically every other nation on Earth. He called it 'an agreement that disadvantages the US to the exclusive benefit of other countries', although which countries he meant was not obvious. In one sense it was tantamount to saying it disadvantaged the United States to the exclusive benefit of the planet. On his first day in office in 2021 President Biden signed an Executive Order committing the US to fully rejoining the Paris Agreement.

All the same, this former loophole is only theoretically closed because it remains voluntary for countries to include their armed forces in their obligation to cut carbon emissions. Unsurprisingly, therefore, huge obfuscation still surrounds the military spending even of NATO countries, let alone that of nations who feel less inclination to be good global citizens and report their militaries' carbon footprint accurately. The US claims its armed forces emit 56 million tons of CO_2e (carbon dioxide equivalent) each year but SGR estimates the true figure is nearer 205 million tons. Britain's Ministry of Defence claims the UK military's total annual carbon footprint is 3 million tons of CO_2e while SGR estimates it is probably 11 million. The largest carbon emissions by any UK-based company happen to be those of BAE Systems, which alone accounts for some 30 per cent of the UK's defence sector's total production.[4]

It is not easy for any citizen to view their armed forces objectively. The military everywhere is largely accepted as guaranteeing its country's continued existence in a potentially hostile world. In view of its often near-sacred status it seems almost churlish – even unpatriotic – to expect it to abide by the same standards and practices as the rest of us. At the same time, most taxpayers see no evidence of how the prodigious sums of money extracted from them for the military each year are being spent, let alone whether they are being spent wisely. The money is simply poured away by the ton into a khaki-coloured hole from which brasshats occasionally emerge talking about 'threats' and lamenting cutbacks before taking early retirement and going straight into well-paid jobs in the defence industry. Easily bought off by the glitz and rainbow smokes of air displays and the tabloids' jingoistic accounts of past military glories, the rest of us go on treating the military as though they inhabit an entirely different planet in which normal morality and accountancy hardly apply: a planet that also magically lacks an environment. This trick really only works because military affairs are so hedged about with laws and restrictions that it is almost impossible for ordinary citizens to obtain certain facts and figures about their own armed forces. The more revealing (or embarrassing) these are, the harder they are to access. Uncovering – and still more divulging – sensitive information is generally considered a threat to national security and is surrounded by thickets of dire threats and penalties.

Nevertheless, since we are here considering an existential threat to the planet and the human race as a whole, matters of atmospheric and other pollution have now become entirely legitimate – even obligatory – subjects for investigation. The more that is discovered about their true extent, the more essential it becomes to pin down some hard facts concerning their causes. It is tempting to look first at the US military as most lending itself to examination, not least because it belongs to a democracy that prides itself on being the most open to public scrutiny and accountability. Many another democracy is not so forthcoming – the UK, for instance, which under the bland catch-all of 'national security' has always been far too ready to invoke the Official Secrets Act merely to save governmental face.

To be fair to the US military, it does not exist entirely in a world of its own and has long been aware of climate change and what it calls the 'threat multiplier' this poses. In fact it has conducted 'think tanks' at the highest level in the Pentagon whose recent official conclusion has been that, always excepting the outside possibility of nuclear war, the gravest threat to the United States is that posed by climate change. (This was before December 2021 when three retired US generals made a statement predicting a second American civil war if a faction in the military refuses to accept the result of the 2024 Presidential Election.) Climate change alone is already contributing the most to global instability by generating such things as large-scale migration away from regions where human habitation is fast becoming impossible. The US itself has been made all too aware of this by ever-increasing high temperatures, fires and devastating tornadoes in its own home territory, even as its southern border is permeated by refugees from all over Latin America. Various coastal bases belonging to all three services have made provision for projected rising sea levels and worsening hurricanes, and here and there virtuous outcroppings of solar panels have sprouted. But as *The Ecologist* has observed, American military policy towards the climate remains highly contradictory. 'There have been attempts to "green" aspects of its operations by increasing renewable electricity generation on bases, but it remains the single largest institutional consumer of hydrocarbons in the world. It has also locked itself into hydrocarbon-based weapons systems for years to come by depending on existing aircraft and warships for open-ended operations.'[5] Even so, it is only fair to note that the US military is conducting a good deal of research into so-called 'Green diesel' – i.e. fuel produced from biomass.

FUEL CONSUMPTION

Recent work reveals that the US military alone burns more liquid fuels and emits more climate-changing gases than do most medium-sized countries.[6] The United States as a whole consumes around 18 million barrels of oil per day, of which the military's share is approximately 1.5 per cent. That represents a daily fuel use equivalent to that of 5 million of its civilians. In 2017 the US military was buying some 269,230 barrels of (crude) oil a day.[7] Crude oil consists of between 83 and 87 per cent carbon by weight. A barrel contains 159 litres of oil weighing around 135 kg of which roughly 118 kg is carbon. Carbon makes up 27.27 per cent of CO_2 gas. If we assume that all the carbon in a barrel of oil will bind to oxygen molecules and make CO_2, then a single barrel of crude could theoretically yield 433 kilograms of carbon dioxide when burned. Thus the US military's daily purchase of crude oil could, when burned, yield 116,576,590 kilograms (or getting on for 115,000 UK tons) of carbon dioxide. Over a year, that would amount to some 43 million tons. However, this figure is still too low because, as noted earlier, the US military admits to 56 million tons of CO_2e while Scientists for Global Responsibility put that figure as closer to 205 million tons a year. Clearly, barrels of crude oil are an equally crude yardstick by which to reckon, but they almost certainly provide a minimum figure while overlooking other sources of CO_2 such as from the use of liquified natural gas. A less scientific way of expressing the US military's annual contribution of carbon dioxide to the world's atmosphere would be 'a heck of a lot'.

'A typical US tank battalion of 348 Abrams tanks consumes 2.3 million litres of fuel per day [one gallon of fuel will move a tank 0.6 of a mile] ... and an F-16 fighter jet uses as much fuel in an hour as a typical vehicle would consume in two years.'[8] The standard calculation is that a single kilo of jet fuel emits 3.16 kilos of CO_2 when burned. This, of course, is equally true for civil aircraft, which is why it is claimed that if the world's civil aviation sector were a country, it would be the eighth-largest emitter of greenhouse gases in the world, at 2 per cent of the human-induced total.[9]

To consider US military aviation alone: once in the air, a B-52H bomber burns 3,500 US gallons of fuel every hour (2,915 Imperial gallons) and carries 139.38 tons of fuel when fully loaded – enough to give it a range of 8,800 miles. In 2017 the USAF purchased $4.9 billions' worth of tax-free fuel and the Navy $2.8 billion, followed by the Army at $947 million and the US Marines at $36 million. Taken all round, the US military carbon footprint is larger than that of any single organisation (other than nation-states) in the world. This makes America's armed forces collectively the planet's forty-seventh largest emitter of greenhouse gases based solely on the fuel they burn.[10]

There are subtleties involved in trying to discover whether certain military vehicles (especially aircraft) might be inherently less energy-efficient than their civilian counterparts simply because they are not designed with economy as a prerequisite. After all, the taxpayer foots the bills. Some revealing if rough-and-ready comparisons might be made between Boeing's long-lived B-52H bomber and the same company's much newer 787-8 Dreamliner. Both are very roughly the same size in terms of wingspan and length. Both have a range of about 8,500 miles. Both have maximum take-off weights of around 220 tons. But the H version of the bomber dates from the early 1960s whereas the Dreamliner first went into airline service in 2011. That half-century made all the difference, especially in terms of engine efficiency. The B-52H has eight engines; the Dreamliner a mere two. The bomber's Pratt & Whitney turbofans have a thrust of 17,000 foot-pounds apiece. The airliner's Rolls-Royce Trent 1000 engines are of a kind known as high-bypass turbofans, each of which has a thrust of 64,000 foot-pounds. Two major differences between these engine types are in fuel efficiency and noise. The Dreamliner can get by with a total fuel capacity of 126,206 litres; the bomber needs 181,610 litres, military engines nearly always being markedly thirstier than civilian types. The lesser noise of the Trent engine is down to its high-bypass design, which accounts for the large diameter. The B-52's eight older-style engines are comparatively slender and are arranged in four pods of two. They are much less fuel efficient as well as very noisy.

These technical comparisons are worth making because they illustrate a major difference between military and civilian hardware: one that holds good for aviation, for naval as opposed to commercial ships, and for land vehicles in general. This is, that military machines are designed with combat performance uppermost in mind rather than the environment or the taxpayer's pocket. In civil aircraft, reducing engine noise has long been a primary consideration that can have a direct effect on sales. No matter how beautiful Concorde itself looked, its engine noise was always a severe commercial problem. It was most economical at altitude and at supersonic speed and this fatally limited its sales potential because nobody wanted it breaking the sound barrier above their country. (Sonic booms are an open-ended gift to citizens hoping to extract a new greenhouse from a government or airline, or claiming compensation for a nervous breakdown.) Concorde was designed for maximum efficiency at Mach 2 in the thin atmosphere at 60,000 feet: some 20,000 feet higher than regular subsonic airliners. In the military sheer performance comes first, and in all but stealth aircraft engine noise is not normally a prime consideration. Yet oddly enough this is where the environment pays an additional penalty in the form of noise pollution.

All engine noise represents lost energy, whether the vehicle is a family car, a motorbike or an F-16. It is inefficient in the sense that no engine is deliberately designed to produce waste noise rather than power. Imagine designing a gigantic amplifier with massive loudspeakers whose sole purpose was to reproduce the same deafening volume of sound that fills the sky from horizon to horizon when a fighter jet makes a fast pass and leaps into a vertical climb. It would take immense power to duplicate those soundwaves, but such is the energy that fighter jets waste every minute they are in the air. Of course military jet engines are designed for a reasonable degree of fuel efficiency; it is just that by-products such as noise are not normally a consideration. Nor is energy lost as heat; although when the use of heat is deliberate (as in the case of afterburners lit to produce brief bursts of exceptional power for critical manoeuvres such as take-off) extravagant quantities of fuel are also burned that give off equally extravagant amounts of sound and carbon.

SONIC PENALTY

That noise alone entails an environmental penalty has been borne out by several studies that reveal the dawn chorus of birds in cities like London has been getting progressively earlier in order for the wretched birds to make themselves heard above the sheer volume of traffic noise. It is also increasingly apparent that the various loud underwater noises (not all of them explosions) that militaries like the US and Russian navies make when testing weapons or 'stealth' submarine designs can severely damage the acute hearing of marine creatures like whales and dolphins over hundreds of miles, let alone the nervous systems of fish in the immediate vicinity.

WAR

It is certainly worth making the point that war itself hugely multiplies the atmospheric and environmental burdens normally left by the military each day. Not every conflict results in scenes like those in Iraq when Saddam Hussein's troops set fire to oil wells and turned day to night over large areas of the Middle East while also polluting large areas of sea. But war enormously increases the amount of carbon that military forces produce. In the military engagements in Iraq and Afghanistan in 2016 alone the US Air Force consumed more jet fuel (roughly 9.85 billion litres or 2,166,739,991 UK gallons) than all the petrol used by all US aircraft during the entire Second World War.[11] The United States has effectively been at war continuously since 1945, whether openly or clandestinely around the world, and this will have accounted for huge additional environmental and atmospheric costs.

CONSTRUCTION

It is not just arming the world's military that takes its toll on the global environment. Its constructions (whether airfields, ports, townships and the rest) also rack up a big debt. A single main runway designed to take the heaviest and biggest transport aircraft might be as long as five and a half kilometres, up to eighty metres wide and fifty centimetres thick, although the thickness might be increased for the touchdown areas. Most runways are half those dimensions but still call for an awful lot of concrete. The world's cement/concrete industries account for some 8 per cent of global CO_2 emissions. Making the over 4 billion tons of cement now produced globally each year releases over 1.5 billion tons of carbon dioxide.

An additional problem with concrete is that it is made with sand. Unfortunately, desert sand is no use for producing concrete with any tensile strength: the grains are too smooth. The ideal is beach sand, and today sand has become one of the most smuggled and black-marketed commodities. Whole islands are stolen if they are remote enough. Over these last twenty years the Chinese military has been building up formerly uninhabitable coral reefs in the South China Sea with dredged sand for laying military runways and port facilities. Yet China has civilian island projects in addition to military ones and Chek Lap Kok, the artificial island twice the size of Gibraltar that is home to Hong Kong's new airport. One of these, The Ocean Flower, is one and a half times as big as Dubai's Palm

Jumeirah, itself a grotesque vanity project. When complete, The Ocean Flower island will house twenty-eight museums (!), fifty-eight hotels, seven 'folklore performance squares' and the world's largest conference centre. A single statistic will serve to illustrate the scale of China's construction projects. In just three years between 2011 and 2013, China used more concrete than the United States did in the entire twentieth century.[12]

AGENT ORANGE AND WORSE

Military forces often leave behind far more environmentally damaging things than burnt hydrocarbons and atmospheric CO_2 and NOx. The use by US forces of depleted uranium (DU) in such weapons as antitank shells has left its radioactive signature in Iraq, for example, and not only on the battlefield. That country has seen acute rises in birth defects, infant mortality, child leukaemia and cancer in general.[13]

The likelihood of DU being one of the causes of the mysterious Gulf War syndrome that continues to afflict ex-combatants has often been cited. Health problems similar to those suffered by Iraqi citizens have been found in the Balkans and particularly in Kosovo, where DU ammunition was used by both the US and the UK. The desert ranges in the United States where regular training exercises are flown against ground targets using live ammunition must by now be a hazard to health even when there are no aircraft in the sky. The USAF's A-10 'Warthog' ground attack aircraft has the heaviest automatic cannon ever mounted in an aircraft: a seven-barrelled brute of a gun that fires armour-piercing DU rounds at a rate of 3,900 per minute: some sixty-five per second. There must be tons of depleted uranium littering the desert. However, on 27 January 2022 the CEO of Northrop Grumman announced that they will stop production of their depleted uranium anti-tank round, M829A4, within a year. Northrop Grumman's DU supplier is Aerojet Rocketdyne and it was not clear what they would do. The implication is that DU weapons, like cluster bombs and landmines, may now be unacceptable to some investors.[14]

But warfare is not always aimed directly at opposing forces. It often involves attacking the environment itself in order to 'interdict' it to enemy troops. A famous example of this was Operation Ranch Hand in the Vietnam War, when between 1961 and 1971 assorted herbicides and defoliants were sprayed on farmland and forests in order to deny food and cover to the Viet Cong. This had a history, in that during the Second World War British and American scientists collaborated in a search to find herbicides that could be sprayed from the air to kill crops. During the Malayan Emergency between 1948 and 1960 the British combined two of these into a defoliant known as Trioxone. This became the direct inspiration for Ranch Hand in the Vietnam War, when the target was mainly the Ho Chi Minh Trail that wound through heavily forested and mountainous country. The Trioxone mix was later known as Agent Orange, itself merely one of a spectrum of related

chemicals that constituted the 'Rainbow Herbicides', so-
called from the colour-coding on their drums (Agents
Green, Pink, Purple, Blue, White and Orange I, II, III and
Super Orange). They were almost all contaminated by
TCDD, also known as dioxin, a recognised and persistent
carcinogen. It is now more than half a century since the last
spraying sorties under Operation Ranch Hand were carried
out over Vietnam, and although the forest cover is slowly
growing back the residues of the defoliants are still causing
birth defects and cancers.

In all, nearly 12 million gallons of Agent Orange were
sprayed; and those interested might glance at the Wikipedia
entry for Rainbow Herbicides where at the time of writing
there is a photo of rusting drums of these chemicals in long
stacks stretching across Johnston Atoll in 1976. This has a
particular resonance for me, and not merely because I recall
the notice outside the door of the Ranch Hand office on
Tan Son Nhut Air Base, Saigon (as was), that read: 'Only
You Can Prevent Forests'. In December 1990 I was aboard
a British survey vessel chartered by the United States
Geological Survey that was doing sonar scans to map the
sea bed contained by the US 200-mile exclusive economic
zone (EEZ). We were scanning the area around Hawaii that
included Johnston Atoll. However, we were told the atoll
would have to be mapped at a later date because at that
time it was a high-risk, high-security area. This was where,
under the terms of a treaty, the US was destroying its stocks
of nerve gas shells and other chemical weapons from its
bases in Germany. Also being destroyed were those barrels

of Rainbow Herbicides in the Wikipedia picture. We learned that even authorised people with security clearance were not allowed to land if they had beards because their protective masks would not make an airtight fit. Our ship made a point of keeping well upwind at all times.

The ancient adage claiming that 'All's fair in love and war' is not an incentive to bad behaviour so much as a rueful admission that there are no limits to which lovers and combatants may not go. This is, after all, what belated treaties try to curb by outlawing such things as germ warfare. At least efforts have been made to hold various militaries to account for strewing live munitions across foreign lands, planting minefields and using cluster munitions that scatter unexploded bomblets over a wide area that later maim or kill people and animals indiscriminately. One hundred and ten states have now signed the Convention banning the use of cluster munitions, including the UK – which last dropped cluster bombs in the Falklands War and then in Yugoslavia in 1999. The United States and Russia have not yet signed the Convention and it has been alleged that Russia has used them in its war in Ukraine.

As a footnote to unexploded munitions, the granddaddy of them all is surely one of the twenty-six huge 'mines' formed of tons of explosives secretly amassed by British troops beneath German trenches at Messines in Belgium in 1917. Of these, twenty were detonated simultaneously in what was probably the biggest manmade explosion of the pre-nuclear period and killed 10,000 men at a stroke. It was audible in London and even in Dublin. The other six mines failed to explode and over the years their exact positions became lost. One blew up in 1955 as the result of a lightning strike, luckily without casualties. Of the five remaining, one is thought to lie somewhere beneath a barn belonging to La Petite Douve Farm and could explode at any time. When interviewed, the farmer himself is often described as 'phlegmatic'.

ROCKETS AND MISSILES

There is one remaining area of technology that until recently was entirely the province of military and quasi-military organisations, even though much of it was in the hands of civilian science. This is rocketry, whether used for missiles or moon landings. At the time of the Cold War it was largely impossible to investigate the likely harmful side-effects of the fuels used. Since then, with the rise of companies like Elon Musk's SpaceX, Richard Branson's Virgin Galactic and Jeff Bezos's Blue Origin and their suborbital 'space tourism' jaunts, a new era of rocketry has dawned. This has brought welcome news in terms of the propellants used. Blue Origin's New Shepard produces only water in its exhaust since it is powered by hydrogen and oxygen. Other rockets use liquid oxygen and liquid methane. The new generation of space pioneers, not being directly under the aegis of the military, shows signs of being rather more environmentally responsible.

We reserve our real scepticism for those nations intent on the ever-expanding and competitive militarisation of space.

This is because their record back home on Earth has been the very opposite of Green. The fallout in Kazakhstan of propellants like perchlorate and dimethylhydrazine (UDMH) around the former USSR's vast Baikonur Cosmodrome rocket launch site has long produced health problems in the surrounding hundreds of square miles. Even successful launches can produce quantities of both chemicals, but catastrophic crashes of rockets shortly after take-off – as in the case of recent Proton-M failures – can scatter up to 500 tons of poisonous fuel. UDMH in particular is very caustic, a neurotoxin and probably carcinogenic as well. Even in Florida perchlorate has been found in water supplies and local lettuces, traced to rocket launches from Cape Canaveral on the coast. The chemical is used as the oxidiser in solid-propellant rockets as well as in missile fuels and fireworks. As a health hazard it principally affects the thyroid and lungs and most of what has been discovered in water in the United States has been traced directly to military bases or plants operated by federal defence contractors. Such things are equally true in Russia, where several military testing grounds and storage areas for rocket fuel in the Kamchatka Peninsula show clear evidence of leaks. Mortality among local marine life is plain for all to see, and scuba divers and snorkellers complain of stinging eyes and skin irritation. And this takes no account of the abandoned nuclear reactors of decommissioned ex-Soviet submarines rotting away in remote creeks.

Thus it is that the military everywhere, not falling easily beneath public scrutiny, causes some of the grossest pollution on the planet, although always in the name of national defence. At least it is now clear that it is the planet itself we ought to be defending, not a bunch of nation states with the ethos of 'every man for himself'. But it is wearying to go on saying so.

Happy Holidays:
Eco-lodges and Cruises

Nobody has the right to mock anyone for treating a holiday as a reward rather than as an adjunct to their moral or intellectual betterment. They, too, have a fantasy. They yearn to believe that a backlog of cheerless months spent in grey northern Europe can at least be assuaged by a couple of weeks' shameless self-indulgence on a beach beside warm blue seas and beneath gentian skies, just like in the advertisements. Where is the harm in that? Whether or not they also entertain dreams of dalliance with sun-bronzed strangers, their fantasies will probably include a constant supply of tall, cold glasses of exotic drinks topped with little paper umbrellas as a prelude to generous lunches and dinners. These should be precious days for banishing worries about fuel bills, mortgages, school uniforms, a new strain of Covid and endlessly deferred NHS appointments; for totally ignoring stories of impending war, rising sea levels, animal species nearing extinction... and on and on. Now is the time for some hard-won enjoyment, cost what it may, and the environment can damn well carry on looking after itself. After all, if people didn't want plastic finding its way into the oceans they wouldn't go on selling sunscreen in squeezy tubes and packing things in polystyrene beads whose static electricity sticks them to the cat's fur.

Likewise, no one should disparage anybody else's reasonable desire to have a holiday that leaves a minimally polluting footprint. After all, a lot of principled folk have the same ambition for their everyday lives as for their holidays. Given the Church of Greenery's pulpit harangues and forecasts of famine, fire and floods this is no longer even especially praiseworthy. At the very least it's sensible and in tune with the spirit of the age. The same goes for wishing to see something of those rapidly shrinking areas of the globe that are still free from urban bustle and the toll it takes on the body and spirit. Hurry, therefore, while stocks last! There are various opportunities on offer.

ECO-LODGES

'Ecotourism' or 'Green' tourism has for some years been a growth industry. It particularly appeals to better-heeled tourists who like to think of themselves as Responsible Travellers, 'responsibility' having become the resonant epithet that distinguishes them from their irresponsible predecessors. (Me, I suppose, wandering the vast untouristed parts of a quite different planet on a shoestring in the late 1950s and 60s.) Modern Green travellers may now choose to base their holidays in an eco-hotel, an eco-resort or an eco-lodge.

There are some distinctions between these virtuous alternatives. Eco-hotels may well be in a town and usually describe themselves as pledged to be more sustainable. (Commercial enterprises often 'pledge' themselves. It suggests serious intent but falls carefully short of being a legally enforceable promise.) In an eco-hotel this usually means the management congratulates itself on offering such things as energy-saving lightbulbs, water-saving lavatory cisterns and less frequent laundering of bathroom towels, while shunning disposable plastics and avoiding unnecessary lighting by having timer switches in the corridors. All these virtuous measures are, of course, saving the management money. If the hotel boasts a restaurant there will be plenty of provision for vegetarians and vegans. Such things have anyway long since become commonplace in the hotels and public buildings in most tourists' home countries, if not in their own homes.

Eco-resorts, on the other hand, tend to be Green hotels sited out of town, often on islands or in mountains. Being more than merely somewhere to spend the night Responsibly, they usually offer such amenities as beaches, spas, infinity pools and a choice of restaurants, most likely together with tours to examine the local flora and fauna. Much will be made of the resort's minimal ecological footprint with emphasis on its own spring water supply, its power derived from solar panels and all the hotel's sewage being treated in natural digesters using an aerobic process free of chemicals or filters.

Lastly, and heading the virtue list, are eco-lodges. These tend to be still more remotely sited and may offer accommodation pared back until it borders on the spartan. They merit closer scrutiny. For a start, why 'lodge'? Well, it sounds less commercial than a hotel, less rowdy and institutionalised and cheapo than a hostel, and a good deal more rustic than a motel. True, one of America's oldest hotel chains is called Travelodge. But with 'eco' rather than 'travel' tacked on front, 'lodge' manages to sound bespoke yet faintly ethnic: smaller than a caravanserai and less fatal than a hospice. www.tourismteacher.com has this to say about eco-lodges:

> They are usually small, generally with fewer than 30 rooms. This means they always have less impact on the environment. You will find eco-lodges located in a natural area, or in a rural area within a short distance of a natural area. They tend not to be significantly impacted by a townsite, noise, traffic, smog or pollution.

This merits deconstruction. Having fewer than thirty rooms says nothing whatever about their potential environmental impact. A twenty-room super-exclusive hotel down the coast from Lagos might offhandedly be doing its daily bit to wipe out the local mangroves. The woolly distinction of a 'natural' as opposed to a 'rural' area is also typical of this sort of copywriting, which generally interchanges feelgood adjectives like 'natural', 'rural' and 'pristine' with more abandon than accuracy. What exactly would an unnatural area be?

'Eco-lodges will contribute to the local economy. They help demonstrate that ecotourism is a more sustainable long-term way to earn income than by destroying or changing local habitats for short-term capitalist gains.'

Ah, this is a message to the natives. Ignoring the fact that any hotel must inevitably contribute something to the local economy if only by employing cleaners, we note the airy disparagement of 'short-term capitalist gains' to describe Third World survival without explaining the pressures of Third World living. The implication is that without the eco-lodge's example the locals would be feckless and too easily seduced by foreigners' money. My own years of living in a South-East Asian subsistence fishing community brought it home to me daily that the villagers, not being stupid, would be only too glad to be able to get by without having to damage their own local habitat. They are perfectly aware that allowing ruthless Chinese traders to strip-mine their corals of tiny colourful fish for the international aquarium trade in exchange for minimal cash is cutting their own throats. Local boys are recruited to swim down with plastic bottles and squirt weak cyanide solution into the corals, and before long the stunned fish start floating out of the holes and crevices. In a boat overhead more boys with small nets scoop out those not obviously dead and put them into a large plastic bag of seawater into which oxygen is bubbled from a rusty cylinder. Few of the poor creatures will have survived poisoning and trauma to be still alive by evening, and even fewer will outlive the ordeal of international air freight in order to spend the remainder of their lives staring through glass at the waiting room of a Los Angeles plastic surgeon.

The point is that everyone in the village is perfectly aware of the dangers of depleting and poisoning their reefs, but the demands of daily survival are paramount. For people with no financial cushion life is entirely a matter of short-term exigencies. The jeep fare to town for essential shopping, the bottle of medicine, the nylon twine needed to repair nets, the school exercise book, the pair of flip-flops, the kilo of rice: all demand cash today, so any gain is better than none, no matter how regretfully it might have to be made. Yet this is still not to identify this paragraph's subtle message, which is that the locals ought to realise how ecotourism is doing them a favour by offering 'a long-term way to earn income'.

Lodges aim to inform guests, staff and visitors on the importance and value of a healthy ecosystem. They will show how to best enjoy the area without impacting it. Often, eco-lodges offer programs and activities that allow guests to work on the land and give back to the local community too! Some will have farms on site, and some will offer you the chance to work.

Holidaying in an eco-lodge that provides its like-minded guests with opportunities to work (magically without 'impacting' the place) sounds as though it might be more like visiting a kibbutz than dropping into a Club Med, but each to his own. In any case the lodges' message to the locals is plain enough: It will pay you to see your future as ministering to principled rich foreigners who have travelled thousands of miles to play for a fortnight at aping your livelihood by learning rubber tapping, harvesting mealies or planting sorghum. They are not surviving; they are missionaries for Greenness who have paid heavily to come to the place where, willy-nilly, you and your family were born and live. Frankly, you're not likely to learn anything useful from them. Your motto should definitely be: Never mind being Green: concentrate on the Greenbacks and those long-term capitalist gains!

The irony arises from the implication that subsistence living in 'natural' areas doesn't also take its toll in harder and shorter lives. Tropical idylls only ever exist in the inner eye of beholders nurtured on Hollywood, Gauguin's Tahitian paintings and David Attenborough's documentaries. Wishfulness edits out the harshness, disease and corrupt local officials. Surely what is left must be more natural: more unspoilt, less contaminated?

The MahaRaja Eco Dive Lodge in Raja Ampat, Indonesia, is likewise the soul of Green virtue although perhaps aware of needing to reassure guests:

> There's no running water, but it's far from roughing it. Fresh spring water is used for the Western-style showers and toilets and the guest rooms are supplied with biodegradable toiletries, corn-based toothbrushes and mineral-friendly SPF. Vegan renditions of traditional Papuan meals are made with locally sourced produce.[1]

The copywriter's art is apparent in the phrase 'fresh' spring water – as opposed, presumably, to stale spring water. If there are showers and toilets but no running water we assume servants with buckets regularly top up the cisterns: 'Roughing it' would probably imply guests having to do that themselves. As for the reference to the sunscreen's protection factor, the phrase 'mineral-friendly SPF' makes no scientific or any kind of sense (exactly how does one show friendliness to minerals, or they to you?). It presumably refers to creams that incorporate titanium dioxide (TiO_2) and zinc oxide (ZnO), i.e. physical sun-barriers that aren't absorbed into the skin. On the other hand they are no less 'chemicals' than avobenzone ($C_{20}H_{22}O_3$) or octinoxate ($C_{18}H_{26}O_3$), both sunscreens that are absorbed. Given that any substance on Earth – especially a mineral – can be assigned a chemical formula, it doesn't seem a useful distinction to make except that it takes for granted eco-guests will already know that avobenzone breaks down quickly in sunlight so doesn't offer much protection as a sunscreen after an hour or so, while octinoxate has come under suspicion as possibly damaging corals. All the same, 'chemicals' has now become one of those damning words in Green vocabulary. Anything 'friendly', on the other hand, sounds cuddly; so 'mineral-friendly' evidently means that the elements titanium and zinc are Good while other elements like carbon, hydrogen and oxygen are Bad. (But of course carbon is Bad. We all know that.) Those who instinctively shy away from 'chemicals' would undoubtedly suspect the very worst from $C_{55}H_{72}O_5N_4Mg$. Just *look* at all that carbon! In fact it is the formula of the commonest kind of chlorophyll: the green pigment essential for photosynthesis, without which plants wouldn't be green and environmentalists would need a new label.

All the same, we suspect that one or two of the guests will have smuggled the odd miniature aerosol of mosquito-killer in their handbag. Life in the tropics can be made a misery by insects and it only takes one trapped inside a mosquito net and whining about one's ears to make sleep impossible. Anyway, it's a very small aerosol and they can't believe it's going to make any serious impact on all that jungle outside.

As for those locally sourced vegan renditions of 'traditional Papuan meals', it is cheering to remember the ancient Papuan tradition of ritual cannibalism that from recent evidence still lingers here and there. This would present an interesting culinary challenge to any Vegan chef worth their salt, even as wandering the local forest could offer an unforgettable experience for a plump ecotourist.

Granted, much of the above deconstruction is quibbling and nit-picking. But it is the awful virtuousness of Green tourism and its clichés that drives one to it. The eco-lodges; the 'giving back to the community'; the horror of 'short-term capitalist gains' that adroitly sidesteps a whole thorny world of politics and economics in underdeveloped regions. Above all, and finally to put things properly into perspective, we need to remind ourselves that in order to be suitably idyllic, eco-lodges need to be fairly remote. They are typically sited in Africa (especially The Gambia), Indonesia and Costa Rica. In other words, in order to holiday in one of them most Green tourists will need to fly several thousand miles there and back, leaving behind them many a black carbon footprint in the upper atmosphere. Moreover, these journeys will inevitably be topped and tailed by car, bus or jeep rides. The remoter the lodge, the less likely the vehicles are to be electrically powered. On arrival, it is all well and good to dine on locally sourced fish and plantains by the light of an oil lamp or just a sputtering castor oil bean ('the poor man's candle'); but even the Greenest of modern tourists – and especially those with children – will usually want some electricity to recharge their lifeline gadgets even if they are prepared to ditch their electric toothbrush in favour of a corn-based one. Rather than a chugging generator hidden in the forest, this probably argues a considerable area of solar panels that won't make the place look any more 'natural' and certainly not pristine.

It is interesting to note certain parallels between www.tourismteacher.com's blurb and similar advertisements for exclusive eco-cruises on small ships to places like the Arctic and the Antarctic. Stress is invariably laid on the eco-friendly nature of the vessel and its rigorously limited carbon footprint. Such cruises are now often sold on the promise of being accompanied by a well-known authority on the area and its animal denizens who will offer after-dinner talks on what passengers can expect to see the following day and answer pertinent questions about walruses, plankton and frigate birds. These experts are the counterparts of the eco-lodges' 'interpretive nature guides' who can 'provide education for guests and visitors'. Even so, carbon is not the only kind of footprint an 'eco-friendly' vessel can leave. This was illustrated in December 2021 when scientists from the University of Basel were in Antarctica's Weddell

Sea researching microplastics in seawater. They found that nearly half of all the samples they analysed were from marine paint. Then to their dismay they discovered that 89 per cent of these micro-fragments were paint off their own vessel, RV *Polarstern*, showing that even when we think we are tiptoeing Responsibly in a wilderness we are still leaving footprints indelibly behind in the form of microplastics abraded from out hi-tech bootsoles.[2]

The idea that Green tourism should be at least partly educational rather than purely hedonistic has a long history dating back at least 2,000 years to wealthy Athenians' tours of pharaonic Egypt. (Graffiti scratched on the Pyramids and other monuments testify eloquently to their children's ennui.) The same notion was popular in Europe between the sixteenth and nineteenth centuries in the form of the Grand Tour that included legendary landscapes and famous artistic cities like Florence and Venice. In those days everyone accepted that travel was hard work at best and mortally dangerous at worst, but that culture and novelty were its own reward. At the very least one faced unmade roads, bandits, greedy local officials and uncomfortable horse-drawn coaches that periodically lost a wheel. Then there were the inns whose beds were crawling with lice or already full, obliging male travellers to sleep on straw in the stables. Thomas Cook himself would have recognised today's eco-lodges as filling exactly the same function as some of his early outposts in painterly locales such as Mont Blanc, Capri and Pompeii. Entire local economies took shape around catering for these foreign visitors. New townships sprang up to supply the amenities the tourists felt they needed once they had paid for the romantic experience by roughing it on the journey.

Green tourists should take heart from history. When the motor launch taking them to their eco-lodge breaks down in a remote and unspoiled tropical cove, they should remind themselves that it is part of an ancient tradition: merely today's version of having one's postilion struck by lightning. Like their historical forebears, the tourists can reflect philosophically that it's all a bit of an adventure in the name of broadening their experience, even as clouds of mosquitoes home in on them while they bob helplessly on the sea as the tropical dusk falls. Everybody agrees not to inform on the lady with the blessed aerosol of insect-killer. In due course when they are safely back home in the suburbs thousands of miles away the jellyfish stings, amoebic dysentery and infected monkey-bites will have turned into badges of Green honour.

It is probably just as well that most holidaymakers don't want to book eco-lodges in remote and underdeveloped rural areas. If they did, yet more areas would steadily become developed in all the ways Green travellers don't want, and they in turn would have to move further into the ever-shrinking wilderness for their holidays.

CRUISE HOLIDAYS

Holidays taken in huge cruise liners represent the opposite end of the scale to eco-cruises and eco-lodges. Again, this is not by any means a new industry. The idea of ship-borne tourism goes back to just before the dawn of the Victorian era. In 1835 the editor of *The Shetland Journal* proposed organising summer cruises to the Faroe Islands and Iceland, and in winter trips to sunny Spain. In 1840 the Peninsular & Oriental (P&O) Line offered a trip to the Mediterranean for forty passengers, one of whom was the writer William Makepeace Thackeray who afterwards wrote a series of light articles about it for *Punch*. Five years later a notice appeared in a Leipzig magazine for 'An opportunity to take part in a world cruise.' All these voyages took place in steamships that had been designed to take goods and passengers to destinations around the world for business rather than for pleasure: that was how overseas trade went on and empires were run. In those days sea travel was still risky enough to be undertaken only by those who couldn't avoid it. Steam power was still supplemented by sail so that all might not be lost if the engines broke down, as they frequently did. This made any passengers going purely as sightseers quite brave and enterprising. In those pre-radio days ships frequently disappeared without trace and everybody grew up knowing that sea travel was lethally unpredictable.

In October 1858 P&O's steamship *Ceylon* sailed on her maiden voyage from Southampton to Alexandria via Venice and other Mediterranean ports with a handful of passengers who probably considered themselves at least as intrepid as their coach-borne counterparts. The ship had been designed for much longer voyages to India, Ceylon and Australia with provision for 130 First Class and 30 Second Class paying passengers, and this initial trip to the Mediterranean would have been an ingenious way of making the trial run of the new vessel pay for itself. Later, after more than twenty years' service the *Ceylon* was re-engined and refitted expressly for cruising rather than for cargo. The following notice appeared in the *Daily News* of 17 September 1881 to advertise the first-ever around-the-world cruise in the world's first purpose-built cruise ship:

> The CEYLON is now lying in Victoria Dock where she may be seen by any persons who have an inclination to voyage in her. Her arrangements are all that can be desired. Beside the usual accommodation on vessels of her class a luxurious boudoir for the exclusive use of ladies, and a capital smoking room for gentlemen, are erected on the upper deck, while the berths are extremely convenient, two persons only being allotted to each cabin. The voyage will be commenced on October 15th and will terminate about July 7th, 1882, Captain R. D. Lunham commanding. Every accommodation has been made for conveying private servants of passengers and the vessel is so arranged that the latter will have the whole of the main deck to themselves, without being interfered with by the crew in any way. The charge for the entire cruise is £500 and £150 for the passengers and their servants respectively. Should the present speculation turn out a success, it is the Intention of the Company to organise a regular series to such entertaining and Instructive voyages.

At today's values that could amount to very roughly £60,000 a head for nearly nine months' 'entertaining and instructive' cruising, plus another couple of thousand if you took Jeeves along with you. At the end you might rightly claim you had made history by having sailed around the world for no reason but the experience itself. Not even Magellan could claim that.

For most of the next hundred years many of the ever-grander transatlantic and long-distance liners doubled as cruise liners according to season, but by the 1930s air travel was beginning to eat into their profits. This process accelerated after the Second World War with its great technological strides in aviation, and by the 1960s transatlantic liners had become dinosaurs. At the same time flourishing economies now entailed higher wages and statutory holidays for potential travellers. The idea of getting away from it all aboard a floating hotel that could offer more than home comfort and a huge array of diverting things to do led to today's generation of purpose-built cruise liners.

These have now carried to an extreme something the old passenger liners had increasingly been designed to do. This was to distract passengers from being aware they were aboard a speck in the ocean, held by a few inches of steel above countless fathoms of freezing dark seawater that, if something went calamitously wrong, would swallow them up without so much as a belch, just as it had the *Titanic* and countless other vessels. The old liners were increasingly set-dressed so that if they didn't want to, passengers never needed to go up on deck and brave the wind and spray and smuts from the funnels. Instead they could relax in a perfect shipboard facsimile of the drawing room of a grand country house, complete with curtained French windows and a blazing log fire; or else they could be in a fabulous hotel or palace with a sweeping marble staircase down which one expected Ginger Rogers and Fred Astaire to tap-dance at any moment. Swimming baths with faux-Grecian columns were incorporated in liners even before the First World War and the baths themselves became a fixture as essential as a choice of dining rooms.

Today's cruise ships have taken this principle and elaborated it to the point of becoming floating fantasy-lands of stupendous size and resources. To anyone brought up on the rakish elegance of those 'greyhounds of the ocean' that were the last transatlantic liners, modern cruise ships are mostly fairly hideous, resembling floating blocks of flats. They may be further disfigured by graffiti-like slogans or dreadful artwork such as on AIDA Cruises' *AIDAprima* that sports a huge eye painted on each side and red lips across the bows. It is curious to note how similar to one another cruise ships and container ships are beginning to look: the one with sheer cliffs of cabin windows (not portholes); the other with towering stacks of units locked together like Lego bricks. But there, it's all freight. The latest cruise ships are some of the biggest vessels afloat, in raw dimensions exceeded only by a few bulk oil carriers. Royal Caribbean's latest 'Quantum-class' cruise ship, *Odyssey of the Seas*, is bigger than the world's biggest, newest and most expensive warship, the USS *Gerald R. Ford* aircraft carrier that has upwards of seventy-five aircraft aboard and a flight deck 330 metres long. That is staffed by some 4,400 personnel. The *Odyssey* has provision for 5,400 passengers even before the crew is counted. In fact, today's largest cruise ships are essentially mobile cities that provide far more amenities than most towns with twice their population. Almost the only way in which the aircraft carrier exceeds the cruise ship is in its cost: roughly $12 billion as opposed to around $1.43 billion. This merely reflects the difference between state-of-the-art military hardware charged to the taxpayer and a floating hotel that has to return investment.

Nothing better illustrates the difference between Thomas Cook's original tourists and those of today than the pictures of a modern cruise ship moored in Venice and dwarfing it. The grandest Renaissance churches and palaces are reduced to looking like miniature buildings on a model railway layout, the Grand Canal shrunk to a rivulet. This is not the place to discuss the fallout of these ships in the ports in which they are so loved and so hated. Thanks to decades of civic outcry and belated legal action there is now a limit to the amount of exhaust pollution cruise liners are allowed to generate both in port and on the high seas. Also, their propulsion is constantly being revised in view of new technology, and many of the latest liners are powered by liquid petroleum gas (LPG) which, though still a fossil fuel, is a lot cleaner than bunker oil and marine diesel.

But here is an important and almost entirely overlooked part of the environmental cost of travel by whatever means, in coaches or aircraft or cruise liners. We are told that a typical cruise liner will burn 300 tons of fuel a day, with the emphasis on the fuel's carbon footprint: a direct way of taking account of the long-term cost to the climate, just as with cars and aircraft and any other vehicle. But what about the massive ecological cost of building the thing in the first place? Nobody ever mentions that. Similarly, in the case of electric and hybrid cars there are few environmental inventories made of the steel and aluminium mined and smelted in order to make them or the rubber and plastic, the nickel and rare earths for the battery, the copper wiring and the rest, right down to those six-year-old African children labouring in Chinese cobalt mines. All we usually hear about is the environmental cost of running the things. For an airliner such construction costs have to include sophisticated materials to withstand the stresses of high speeds and big temperature ranges. But building a floating city (which is the modern cruise ship and its contents) entails a prodigious environmental debt in terms of the materials sourced and used. Yet who ever reckons it? A new cruise ship might have upwards of 3,000 cabins: roughly equivalent in size to a big hotel in Las Vegas capable of accommodating over 5,000 guests. God knows how many miles of carpeting alone are required (every inch of it deriving from Big Oil), or how many tons of paint (ditto). And that is without the equipment: the highly sophisticated engines and podded propellers and all the rest of the navigation and technical wizardry that is duplicated and often triplicated in the name of safety. Yet we never hear of these ships' total construction and contents being itemised and ecologically costed, only (if at all) their estimated footprint when sailing.

So how does that massive environmental debt get amortised or sequestered? Answer, of course: it doesn't.

The Covid pandemic famously dealt a death blow to many forms of holidaymaking and many a perfectly good – even nearly new – cruise liner was towed to breakers' yards in Bangladesh and elsewhere and demolished to save on mooring fees, skeleton crews and the fuel needed to run auxiliary engines to keep vital systems alive. But cruising now occupies an economic niche too big and powerful to be allowed to fail, and new ships were being built even as people emerged, pallid from lockdown, eager to indulge themselves with the sort of holidays that only cruising can give. Modern cruise liners are expected to offer almost anything pleasure-seekers could get on dry land but might have to drive miles to find. Here are pools with artificial waves and surf, simulated skydiving, rock-climbing walls, solariums, outdoor movies, jogging tracks, saunas, gyms, roller-skating rinks, ice-skating rinks, dodgems, fitness centres, children's crèches and playrooms, dozens of bars and up to twenty restaurants catering for any known cuisine and several much better left undiscovered. Plus of course a high street containing the sort of shops – or rather 'outlets' – that clutter up airport terminals selling the usual name-branded stuff. God forbid that people wanting to 'get away from it all' would have to give up shopping.

And all of it in motion. Periodically the travelling city stops at a famous place with opportunities for sightseeing. It is the last remaining vestige of the old Grand Tour. Cruise passengers are often fleeced once they get off the ship, and for some the sight of a foreign town is almost the first they are aware of having been afloat. Not every hedonist wants to risk having to encounter Culture. P. G. Wodehouse expressed this pungently in *The Code of the Woosters* when Bertie resists Jeeves's hope that they might go on an around-the-world cruise. Among the drawbacks Bertie cites is 'the nuisance of having to go and look at the Taj Mahal'. He would have been entirely on the side of the Greek teenagers who long ago scratched their disaffected messages ('When's lunch?') on the walls of ancient Egyptian temples in Thebes.

Mobile Phones & Computers

Over the last twenty years mobile phones have become almost as essential to most people's lives as their clothes: an inseparable companion, often closer than many a friend, and not infrequently more indispensable even than a spouse or partner. American teenagers have been quoted as believing a person's mobile phone is the most important indicator of their social status or popularity – outranking jewellery, watches and even trainers. The social costs as well as rewards of this portable world have often been examined, but the environmental cost of mobile phones and their service providers is more easily overlooked. As mentioned before, no one likes to think of the environmental consequences of things we deem essential to our lives although it is still not as difficult as admitting the downside of anything we really enjoy (holiday flights, G & T, online shopping, driving fast cars, foie gras, cocaine, whatever).

In the case of mobile phones we can get a good idea of their environmental cost thanks to the work of Mike Berners-Lee, the expert on carbon footprinting and brother of Sir Tim, inventor of the World Wide Web. According to Mike a year's typical usage of your mobile for up to two minutes a day cashes out at forty-seven kilos of CO_2. An hour a day produces one and a quarter tonnes in a year. Over that same year the world's mobiles will account for 125 million tonnes of CO_2. If you are one of those who chat on their mobile for an hour a day, then by the end of the year the amount of CO_2 you will have produced is the equivalent of flying from London to New York in economy class, one-way.[1]

The direct energy cost of charging a mobile phone is minimal since the amount of electricity needed is tiny. It is the hidden, passed-on costs that are of a different order. Each use of a mobile phone that requires making a digital connection consumes very much more power, but thanks to distributed processing on a global scale the electricity is invisibly consumed thousands of miles away. Every time a phone, computer or smart TV is used to access the Internet a request is sent to one of the enormous data centres around the world: air-conditioned complexes that are among the planet's most energy-intensive systems. By some calculations these centres alone absorb approximately 3 per cent of the entire world's generated electricity, an amount that can only rise as Internet-based systems proliferate. Already a single major data centre can easily consume as much electricity as a medium-sized town. One expert claims that 'about 5 per cent of all global energy use is by electronics. That number is projected to grow to at least 40 per cent by 2030 unless there are major advances made in the lowering of electricity consumption.'[2] Which is not going to happen.

In the next few years a likely billion more people in developing countries will be coming online, bringing with them their own high-resolution photos, streamed videos, emails, surveillance camera footage, mobile phones, TVs and 'smart home' devices, all of which will require still more computing power, greater amounts of network bandwidth and ever-greater storage capacity. To the Internet of Things (IoT) you can add artificial intelligence (AI), driverless cars and robots. As Mark Zuckerberg's plans for a 'Metaverse' come to fruition and more people choose to desert reality and spend their days wearing headsets and inhabiting a digital world, the amount of computing (and hence electrical) power demanded by data centres in the real world will vastly increase. The Metaverse and its possible rivals will exist only in its inhabitants' linked computers. Virtual reality worlds already exist and are attracting young fans from around the world. Their 'avatars' (computer-simulated alternate 'selves') are buying real estate and Gucci handbags in their digital shadowlands for which they pay with real currency. This strikes most people as the height of insanity. Mad or not, the energy expenditure is only too real, like the dollars.

Estimates suggest that data centres will become the world's largest users of energy by 2025, and that is over and above the increased need for natural resources such as rare earth elements to make the screens, goggles and processors. Somehow, there seems little point in building a digital world if the physical one burns while we do so.[3]

MANUFACTURE

The environmental cost of actually making all this electronic gadgetry is something else. Even if mobile phones were low-tech objects like cricket balls, their sheer number would have an impact; but the manufacture of the world's estimated total of 6.1 billion mobile phones employs millions of people in dozens of different industries and takes an enormous environmental toll. A team from Canada's McMaster University recently found that mobile phones were the most damaging of all electronic devices in terms of the emissions incurred by their manufacture from start to finish.

Generally speaking, about 40 per cent of a mobile phone consists of metals (predominantly copper, gold, silver, platinum and tungsten), another 40 per cent of the phone is plastic of one kind or another, with the remaining 20 per cent being taken up by various other materials including glass and ceramics. Each phone can contain as many as sixty-two different metals, plus sixteen of the seventeen 'rare earth' metals whose sources are inconveniently scattered about the globe in pockets that are mostly difficult to mine. No one knows the true extent of these metals' reserves but there are no known substitutes for them and once they are gone, they're gone. This makes recycling old mobile phones all the more important. In addition, one of the elements that goes into mobile phones is tantalum, whose source is an ore known as coltan. Tantalum is vital for the world's electronics industries. It is an ingredient in all sorts of devices, both civilian and military (it is essential for the circuitry of smart bombs). Well over half the world's known coltan reserves are in the Democratic Republic of Congo and neighbouring Rwanda. The civil wars in both countries make tantalum a conflict element but, unlike with diamonds, its exact source cannot be identified from analysis.

The fact is that with the mining of various minerals essential to mobile phones and the slavery, deforestation and spills of poisonous slurry associated with that mining, this ubiquitous gadget is doing enormous damage to the planet, both in its manufacture and its daily use. The sheer numbers of this newly essential prop to civilisation constantly exacerbate its impact, and not least because although some degree of recycling has belatedly become a priority, the phone's very miniaturisation makes recovering the more valuable trace elements slow and labour-intensive and it is likely that no more than 30 per cent of a mobile phone can be recouped and used again. The rest simply goes into the world's annual millions of tons of e-waste. It does seem curious that an estimated 4,000 tons of silver, 380 tons of gold and 200,000 tons of copper should be locked up in unused mobile phones scattered all over the world. The Global E-waste Statistics Partnership puts it all into grim perspective. In 2019 a record 53.6 million tons of e-waste was generated worldwide – up 21 per cent in five years. This mountain of once-costly junk includes laptops, monitors, electric toothbrushes, mobile phones, air conditioners, toasters, vibrators, TVs... In short, the entire list of the varied electrical equipment that mediates our lives.

In 2019, only 17.4 per cent of e-waste was officially documented as formally collected and recycled. This means that iron, copper, gold and other high-value, recoverable materials conservatively valued at $57 billion – a sum greater than the gross domestic product of some countries – were mostly dumped or burned rather than being collected for treatment and reuse in that year.[4]

Those 53.6 million tonnes are foreseen as likely to double in the next ten years, much of the increase being down to the goods' shorter lifespans. As with all consumer durables that double as fashion accessories, built-in obsolescence is their most obvious characteristic and the one with the worst environmental consequences. That is entirely down to competing companies trying to induce consumers to buy the latest version of their mobile phone every year or two, mostly on the basis of a gimmick such as yet another camera or else a wrap-round or even foldable screen. This has far more to do with fashion than with capability. Indeed, few things better exemplify the damage done by our entire economic system than this relentless upgrading of electronic consumer items that often includes features most people will never use.

FASHION

This principle of built-in obsolescence is nicely illustrated by the case of the so-called Phoebus cartel of the late 1920s. This was formed by half a dozen of the world's major lightbulb manufacturers such as Osram and GEC who by 1928 were worried that the average lifetime of a domestic lightbulb had risen to 2,800 hours. This, they felt, was far too long to guarantee steady sales. Their technological prowess was undermining their own business. The cartel therefore agreed to research ways of making bulbs' lives shorter and it was decided that member companies would be disciplined and fined if their products lasted too long. By 1934 they had successfully halved the average bulb's lifespan to 1,400 hours and sales were booming. They were aiming at 1,000 hours but never got there. Instead, the Second World War destroyed the cartel by nationalising or restricting raw materials and putting a premium on economy. Ever since then, the idea of the 'planned obsolescence' of consumer products has had an increasingly bad press as people become more environmentally conscious. All the same, in the world of electronic goods it is still very much alive and kicking. Today's mobile phone giants are quite aware that with the help of judiciously chosen influencers the latest fashion of any consumer non-durable has the lifespan of fresh flowers on a grave.

Again, there is nothing very new in this. Car manufacturers have been using the principle of planned obsolescence for at least the last ninety years, bringing out new models annually whose slight design changes make them identifiably of a particular year and letting their customers' fear of being outdone by the Joneses do the rest of the selling for them. Suddenly, though, thanks to environmental concerns and growing social pressures, it may be that the urge to buy a new mobile phone with each new issue could weaken. The availability of Fairphones that are brand new while incorporating recycled materials could make a welcome difference to the market. Most mobile phones have a potential life of five years and may only need a new battery to carry on well beyond that, and the new move to make them more easily repairable should gradually render such things simpler. Even so, the built-in fashion element means that many mobile phone users only keep their phone for eighteen months or two years before dropping it into a drawer and buying the latest model. Unfortunately, obsolescence may also be built in by the phone companies switching over to 5G mobile networks and suddenly rendering millions of older phones useless.

Noting the habit companies have of bringing out new models more as fashion accessories than to provide better performance, the European Environmental Bureau observed that 'extending the life of mobile phones and other electronics by just one year would be the equivalent of taking 2 million cars off the road in terms of CO_2 emissions.' Not only that, but under new EU rules introduced in the summer of 2021 fridges, washing machines and TVs will all now have to be built to last longer, be cheaper to run and easier to repair. These standards will of course apply to any white goods imported into the EU. We will all want to know how that long-overdue and sensible plan works out.

SUPERCOMPUTERS AND QUANTUM COMPUTERS

Modern mobile phones are effectively online computers, and online computing is much more power-hungry than most people realise. It has been estimated that in the US alone 'classical' (i.e., ordinary digital) computers will, by 2040, require more energy than today's entire national grid can supply. Supercomputers are particularly power-hungry. The world's most energy-consuming supercomputer, Tianhe-2 in Guangzhou, draws eighteen megawatts. Eighteen million watts is an awful lot of electricity for one computer: enough to keep 300,000 60W lightbulbs lit. Its successor, Tianhe-3, will need still more.

Even such advanced, high-speed computers still use 'classical', or digital, systems. The next advance is in quantum computing. In digital computers the unit of information, the 'bit', can have a value of either one or zero. Its quantum computing equivalent, the qubit, can be both one and zero at the same time. This is the idea of superposition: where something can exist in multiple states simultaneously (a baffling notion though perhaps not wholly unfamiliar from the behaviour of certain people one knows. See also Schrödinger's famous cat paradox). In order for qubits' power to be realised, they need to be linked together by a process known as entanglement. In theory, each qubit that is added doubles the computer's power. By attaining a level of 'quantum advantage' some years ago, computers like Google's Sycamore seemed to herald a future that would change the very nature of things like cryptography and AI, both of which require massive numbers of calculations to be carried out simultaneously. In theory this will challenge the hitherto unbreakable anonymity of cryptocurrencies' blockchains, not to mention that of security agencies worldwide (the so-called 'quantum apocalypse'). The technology has been shown to work but it is still a very long way from being able to supplant our domestic, 'laptop' level of digital computing.

At the moment only the wealthiest governments, military and intelligence agencies, university research departments and tech giants are able to afford and operate quantum computers. By 2019 Google announced its Sycamore processor had advanced still further and had now reached the next stage, that of 'quantum supremacy': the ability to perform calculations that a 'classical' digital computer cannot, regardless of how 'super' it is or how much time it is allowed. Sycamore had apparently taken three minutes and twenty seconds to solve a problem that would allegedly have taken the world's most powerful 'classical' supercomputer 10,000 years. A sceptic would want to know who or what calculated that nice round figure. Was it a supercomputer modestly estimating its own performance? Are they sure it wasn't 9,008 years? Or was that just Silicon ValleySpeak for 'a heckuva long time'?

Impressive as 'quantum supremacy' may sound (and no matter how good a movie title it would make) quantum computing may have left its infancy, but it is still firmly in its early childhood. The problem the Google team had given Sycamore to calculate was apparently extremely esoteric and had no application in the real world. In May 2021, one of the world's foremost theoreticians of quantum computing, Isaac Chuang, came clean. 'Quantum computing today,' he said at MIT, 'is actually, from a practical standpoint, quite useless other than for generating publicity.'

This has since become arguable, with all sorts of governments and companies putting considerable reserves of money and talent into research. By no means the least of the effort is being made by China's USTC (University of Science and Technology of China) whose quantum laboratory was opened in 2020 at a reported cost of $10 billion. A sum like that is hardly chickenfeed; and in July 2021 USTC announced it had solved a problem 'three orders of magnitude' harder than that tackled by Google's Sycamore. Only two months later it said it had beaten its own record by a further three orders of magnitude. A USTC satellite has been in orbit since 2016 using quantum communications that are presumed to be unhackable. Quantum communications are widely seen, and increasingly used, as the main defence against the constant cyberhacking that is rapidly becoming a global menace and above all for the military everywhere. Quantum communication does not require quantum computing as such, but it does use qubits. The technique relies on quantum key distribution (QKD). Encrypted data are sent using ordinary binary digits but the keys to decrypt the data are transmitted in qubits.

Since there has never yet been a human technological advance without deleterious environmental consequences, it is reasonable to ask what the downside of quantum computing and communications might be. In February 2018 America's Oak Ridge National Laboratory published a report alleging that quantum computers will consume dramatically less energy than current supercomputers. This may or may not turn out to be true, but the usable technology has not yet arrived on any scale. However, in order to function at all, quantum processors need to be shielded from all stray energy that would otherwise disrupt the qubits' delicate quantum states that so easily decohere if subjected to the slightest heat and vibration. Hence they need to be kept at temperature levels – and therefore energy levels – as close as possible to absolute zero in 'floating' rooms isolated from the slightest vibration. Certainly the processor in IBM's quantum computer in its Zürich 'Q-Lab' is kept at a temperature close to -273°C (absolute zero is defined as -273.15°C). This is colder than interstellar space and is achieved by using liquid helium in a process that requires a good deal of power to get it to within this fraction of absolute zero and keep it there. The superconducting this confers means a quantum computer's circuitry has almost zero electrical resistance.

If they are ever to become as practical as 'classical' computers, quantum computers will somehow need to be scaled up, which probably means five out of every six qubits being used solely for correcting errors. Possibly nothing much short of a million-qubit quantum computer will be able to achieve anything useful. Various work-arounds have been proposed but even experts are not confident of being able to produce a quantum computer that isn't either astronomically expensive to buy, impossible to run cheaply or overdemanding of its environment. So even if quantum computing turns out to require less electricity than 'classical' computing to run the calculations, its refrigeration demands on the electricity grid look like being considerable, as do its highly specialised construction and work environment. In one way or another, this could yet prove to be another of those brilliant inventions humanity might well have done far better without.

(But look! There's a jar with a new kind of banana in it! No power on Earth is going to stop the Monkey putting its hand in.)

As an indication of the gap between this exalted level of technology and that of the humbler everyday, most of the world's countries are still a long way from providing reliable nationwide and global mobile phone coverage. In the Philippine province with which I am most familiar, people often congregate on the beach in all weathers in order to get a signal on their mobile phones. Reference has already been made in the Cars & Aircraft chapter to people in the UK having to walk to the tops of local hills for a mobile signal, and I also know a Briton who can only get partially reliable mobile coverage by standing outside his own garage door in all weathers. To achieve even this basic level of coverage requires immense resources in terms of infrastructure (plus building satellites and getting them into the correct orbit). Ever-increasing amounts of electricity will be needed to power the worldwide network of data centres that make such coverage possible.

Wellness & Beauty

The idea of 'Wellness', with its burgeoning paraphernalia of gurus and gyms and diets and costly remedies, is in some ways a perfect expression of late capitalism and of our decadent, self-obsessed era. Gwyneth Paltrow, with her supposedly vagina-scented candles, fabulously expensive toys and sheer silliness is the era's perfect high priestess. (She has, incidentally, been overtaken scentwise by the German company Viværos's VULVA Original, the details of which I am too old-fashioned to pass on.) The underlying belief of Wellness is that all illness and maybe even death itself can be bought off indefinitely by living 'correctly', which means expensively. Even so, it comes as a shock to learn that in 2018 this worldwide industry was collectively valued at $4.5 trillion: i.e., a sum over half as large as the total annual global spending on health ($8.3 trillion, based on WHO data of the same year). As such, the Wellness industry represents well over 5 per cent of the planet's gross domestic product (GDP).

Yet the Covid pandemic has shown that this majority-female movement has aspects that are more sinister than just the better-off sectors of society being happy to spend money on wanting to be well. Once Covid vaccines were developed the old feminist chant of 'My body, my choice' began to take on newly politicised overtones and certain strands of Wellness began converging with Far-Right and QAnon conspiracy theories. Anyone could see how this was not merely aided but enabled and enthusiastically fostered by social media's wilder interactive websites. Anyone could also see how the wellsprings of Wellness were always very much more transatlantic than European. The underlying reason was surely that unlike in Europe, American medical services are virtually all in private hands and can be cripplingly expensive to the individual. (Even Medicare, available principally to those over sixty-five, will only pay half the cost of treatment and those eligible are advised to take out insurance to cover the other half.) A prolonged illness can beggar even an insured middle-class citizen.

This has arguably had two unsurprising consequences. The first is to make the majority of US citizens almost superstitiously health-conscious and terrified of falling ill. The second is that doctors are easily suspected of being in cahoots with Big Pharma in a way that readily suggests conspiracy, especially after the OxyContin scandal with its 'pill mill' doctors eager to prescribe. The issue of Covid vaccination simply brought these strands to the fore at the height of the pandemic. Suddenly, vaccination could be presented as a politically supported pact between Big Pharma and medical practitioners to make fresh fortunes out of public gullibility, ignorance and fear. And this despite the

US government pointing out that Covid vaccines are given free of charge to anyone; but that if the recipient has private health insurance or comes under Medicare those institutions will be billed to pay the person administering the vaccine at a rate of $16.94 for the first shot and $28.39 for the second.

Given that Covid and its variants is a worldwide pandemic that has already taken a grievous toll in terms of fatalities and economic damage, the idea of guaranteeing that any citizen can get vaccinated for free seems entirely proper and the very least that a responsible and wealthy government could do to avert or slow down further social and economic calamity. But this is not at all how conspiracy addicts see the campaign. To them, what else could this be but the individual's sacred mantra of 'My body, my choice' being overridden under the socialist guise of the welfare of the majority? They saw Wellness as being hijacked by creeping communism as part of a dastardly Washington plot to undermine personal freedom and the ideals of American republicanism. Theirs is a deeply mad society.

Such, more or less, has been the line taken by the wackier online contributors to world disinformation. The fact that it has spread to Europe and beyond – with tens of thousands of citizens refusing to be vaccinated against Covid – is a testament to human pig-headedness. The notion of social responsibility, that citizens might have a duty towards themselves and their fellows by not getting ill and taking up valuable space in intensive care, evidently cuts no ice with such people. Nor does the thought that had their parents and grandparents not been automatically vaccinated as infants against the several killer diseases prevalent at that time, they themselves might never have been born. The chances are considerable that many of today's pampered egotists owe their very existence to grandparents and great-grandparents who were long ago vaccinated against smallpox, diphtheria, polio, measles, whooping cough, TB and the rest. One would need to be deeply deluded to view vaccination as the thin end of a communist wedge rather than a sensible measure of public health like clean drinking water and proper sewage treatment.

But there, much of the vast and lucrative Wellness industry rests on foundations of private panic, scientific ignorance and wacky ideology. Pure egotism, in short. From a commercial point of view it is often nearly impossible to separate the Wellness and the Fitness industries. Both are about Me-Me-Me and My Right to be Immortal; and both depend on a degree of vagueness about which domain each inhabits. For instance, a conventional exchange between normal people might have a solicitous aunt asking, 'Are you well, dear?' to which her niece or nephew probably replies, 'Pretty fit, thanks, Auntie. And yourself?' This response may be what aunts are content to hear, but not Wellness and Fitness freaks and the industries that thrive on them: it is far too vague. It deliberately sidesteps the commercial view that has long pretended to the status of quasi-medical diagnosis: 'Pretty fit, huh? That's what you think, Buster! Believe us, you are nothing like so well or so fit as you think you are or could be.' No matter how well you actually feel, and even if you can indeed walk upstairs without wheezing like an asthmatic walrus, this is calculated to induce a feeling of unease. Yet the statements 'I am fit' or 'I'd like to be fit' are meaningless without the qualifying question, 'Fit for what?' Most people's lives automatically confer on them the degree of fitness they need to live them. Even couch potatoes are arguably fit enough for their chosen career. A century ago many people would walk miles each day to school or work or to visit someone in a nearby village, and the chances are if they were labourers they would be very much stronger in a wiry sort of way than today's equivalents who have machines to do the heavy lifting for them. These days people who cycle to work are unquestionably fitter than couch potatoes, but most of them would be unfit to participate in an Iron Man competition and even more so to compete at Olympic level. To do such things you have to be dedicated to the point of obsession, inevitably aided by special dietary supplements.

Body-sculpting for its own sake has become high fashion over the last decade. Influencers in long-sleeve crops and leggings lift weights and preach endlessly about creatine shakes and protein bars, sales of which have recently ballooned to the extent that entire shops can now be dedicated to them in their myriad varieties. The majority of customers for protein bars turn out to be women. Protein can help give definition to muscles and so produce the sought-after image of the feisty, ripped woman of today as she drifts in her athleisure outfits between yoga classes and powerlifting sessions. (Does she have a job? Can she dress like that in the office?) Why should she care that protein bars usually contain palm oil, which is not only high in saturated fat but is a ubiquitous product for which uncounted hectares of Amazonian jungle continue to be sacrificed? (see Appendix 2, Biofuels). Until a few years ago supplementary protein was almost exclusively taken by male bodybuilders who bought it as a powder in great tubs and mixed it into just about everything they ate. Most was wasted since the body can only metabolise twenty or thirty grams of protein at any one time. As with vitamin C, any surplus is simply pissed away, which in the case of expensive protein supplements is literally money down the drain.

But the fashionable siren call of muscular definition easily drowns the dreary nudgings of common sense. 'Get Shredded in Six Weeks!' booms *Men's Health* magazine from a world where men are wammo, hench and tonk. When I was a boy our magazines were seldom without an advertisement for Charles Atlas, the Italian-American bodybuilder who by means of 'dynamic tension' exercises famously converted himself from a ninety-seven-pound weakling into the sort of hunk who no longer got sand kicked into his eyes on beaches. Even in those days the message seemed to be less about getting girls than establishing dominance over other males, which is also the ethos of *Men's Health*. 'Today's über-tonk males wear their six-packs like beautiful, pointless feathers: this is a cosmetic muscularity rather than a functional one,' as Sirin Kale has memorably pointed out.[1] She added that it is very difficult to acquire an abnormally pumped, low-body-fat physique without chemical help. Time was when notions of health meant ingesting as few 'chemicals' as possible; but today such caution has fallen out of fashion in favour of losing one's heavy former self in the sheer abandonment of chemShredding (and quite possibly chemSex as well). Today's urban Narcissus gazes into his gym's pool and knows that other men envy him his washboard abs. Peak fitness and peak self-love hover like a chimera, faintly scented with chlorine.

Whatever level of fitness he or she aspires to, it will be painfully gained and easily lost once more. This is the hook the Fitness industry uses to keep its clients. After months of gym work people will reach a certain peak of fitness which, unless they continue to punish their bodies and numb their minds with repetitive exercises, they will steadily lose and revert to being a normal slob. It is exactly the same process with slimming. Everybody knows this and the whole subject fills most people with anxiety and a degree of self-loathing. Anyone feeling anxious about their own body is a potential Wellness and Fitness client; and such people are more easily induced to spend large sums of money on themselves and their appearance than they are on practically anything else.

GYMS

Among the environmental consequences of Wellness that are easily overlooked are the gyms themselves: those temples of modern Fitness where the sins of urban living are expiated by people in their millions. For all their healthful innocence, gyms leave their environmental mark. For a start, they are typically in quite spacious buildings needing plenty of heating in winter and air conditioning all year round to deal with the aerobic exhalations and exudations of sweaty bodies. They generally use large quantities of water – heated or otherwise – for showers, and much more if there is also a spa or a pool. A great many towels also require laundering. Cardio equipment (rowing machines, treadmills, bicycles, elliptical trainers and anything else requiring kinetic energy) use power and resources to manufacture, as well as a surprising amount of electricity even if they are only on standby. Gyms also generate a good deal of waste in the form of plastic (bottles, etc.) and chemicals for the pools and spas. Add to that the statistic that 90 per cent of all customers get to their gym by car rather than by walking or cycling.

Efforts have been made to counteract some of these drawbacks by introducing the concept of the Green gym. So far, this has simply meant gyms adopting much the same measures as Green hotels, i.e. LED lightbulbs, reduced toilet flushing and so on. The Green gym concept has also meant that companies which manufacture a wide range of cardio equipment (treadmills, cycling and rowing machines) have mostly now redesigned them to use the customer's expended calories to generate electricity. There is a lucrative trend to declare older, non-generating machines outmoded and in their place install similar machines but now expensively re-engineered to incorporate a small dynamo. This is probably quite acceptable where many customers are concerned because although the gym management will faithfully pass on the cost of the new machines in membership fees, the clients themselves will take heart from the thought that they are being Green even as they sweat.

Alas, the amount of electricity such machines can produce is minuscule. Someone working hard on a bicycle or treadmill might just manage to generate enough current to power the screen they are watching, but that would be about all. A bit of thought will confirm this to anyone who has ever ridden a bicycle rigged to power its own lights. The little generator imparts a slight but noticeable load to the pedalling and all it does is light up a couple of 2.5 volt lamps. This is exactly why most cyclists now carry (and wear) brilliant LED lights powered by battery rather than dynamo. A much better idea would be for the management to cover the roof of the gym with solar panels. Even in the frequently overcast UK these could at least help towards heating the water in the showers or the pool or keeping the lights on.

There is also the matter of the fashion sportswear favoured by gym users who want to cut a dash. The proudly less trendy may have the confidence to ignore the scornful glances of the designer-muscles crowd by changing into holey old cotton T-shirts, bleached-out rugger shorts and ancient trainers recently mauled by the family puppy. But as we have seen in the chapter on clothes, Military Carbon, modern stretchy athleisurewear is practically all made of nonbiodegradable synthetic materials originating in the plastics industry – thanks due here to Big Oil – and using processes that consume enormous quantities of water. There is hope, though. The Taiwanese company Singtex now claims to have created a fabric out of coffee grounds. This is rumoured to be good at wicking away sweat and repelling UV radiation. It is not known whether enough of the aroma of coffee lingers to make the wearer wish they had, after all, simply trotted along to the nearest Starbucks with their iPad rather than subject themselves to agony in the gym and scorn in the showers afterwards. Had they done that, they could also have congratulated themselves on doing their bit for Green sportswear without lifting anything heavier than a blonde vanilla latte.

KEEPING MORTALITY AT BAY

The largest entity within the Wellness industry is the Personal Care, Beauty and Anti-Ageing sector that in 2018 was alone valued at $1.083 *trillion*. This is not only where the industry's Paltrows earn their crust but where a host of mainly female doctors (or at least people on Instagram with 'Dr' in front of their names) wield immense influence with their skincare and other products. Phrases like 'hyaluronic serum' are bandied about (a 10 ml bottle of which will set you back $110) by the likes of Dr Dray (Dr Andrea Suarez), Dr Barbara Sturm, Dr Shereene Idriss, Dr Vanita Rattan and the rest. Hyaluronic acid is a moisturising and lubricating carbohydrate that occurs naturally in the skin, joints and eyes and is much promoted by the cosmetics industry on the grounds that if the little you already have does you good, then lots more will give you back the dewy, virginal look you believe your skin had when you were eighteen and before you took to lying on sunbeds. This is all well and good, and harmless enough except in one respect: that the cosmetics and perfume industries themselves exude varieties of pollution so complex their true consequences are still barely identified and may even be unquantifiable. Microbeads that were once ubiquitous in exfoliant facial scrubs and toothpastes were banned in the United States in July 2018 and after that in Britain and Italy; but there are several loopholes through which other microbead-containing products can squeeze, let alone in countries without bans. As with nearly everything else they wind up in the sea and, after a fish dinner, in us as well. But microbeads are merely part of a very much larger and diffuse problem.

THE 'TOXIC COCKTAIL'

According to the BBC's website on 13 May 2022, microplastics had just been found in human blood for the first time, and in 80 per cent of the people tested. They simply went to join all the other persistent organic pollutants (POPs) already identified in people's bloodstream: toxic chemicals that don't easily break down in either the environment or humans. They chiefly congregate in our blood and fatty tissues. Many people today will show traces of DDE: a metabolite of the long-banned insecticide DDT mentioned in the Gardening chapter.

There is now a newer class of pollutants known on account of their persistence as 'Forever Chemicals'. These are the chiefly polyfluoroalkyl substances or PFAS noted in the Fashion Industry chapter. Hundreds of these are used for such purposes as flame retardants, to repel water in clothing or to make 'non-stick' coatings. The Swedish NGO ChemSec now believes that as much as 62 per cent of all the chemicals currently used in the EU are hazardous to health and the environment. Every one of us has hundreds of synthetic chemicals in our blood: the 'toxic cocktail' whose long-term effects nobody yet knows.

It is one thing to contribute unwittingly to the pollution of the ocean by using microbead-containing products, but quite another to contribute to the pollution of one's own endocrine system via cosmetics. Phthalates, and especially diethyl phthalate (DEP), are used as solvents and fixatives in skin cosmetics, hairsprays and scents. Unfortunately, phthalates are endocrine-disrupting chemicals, meaning they mimic hormones. They can interfere with the body's normal hormone signalling and this can lead to anything from altering mood or energy levels to increased rates of breast cancer and birth defects. In short, such endocrine-disrupting chemicals (EDCs) can damage development as well as the reproductive, neurological and immune systems of humans and animals. They can also cause environmental damage via sewage systems when washed away.

The first study of pharmaceutical pollution in rivers outside North America and Western Europe has discovered that in more than a quarter of the places investigated such contaminants pose a threat to environmental and human health. Two of the most-detected substances were carbamazepine (used to treat epilepsy and nerve pain) and metformin (for diabetes). Also found were high concentrations of so-called 'lifestyle consumables' like caffeine, nicotine and paracetamol. These, of course, are in addition to the well-known river and water contaminants of antibiotics and birth-control pills. Since we increasingly rely on pharmacological solutions to any illness, whether mental or physical, such pollution is only going to get worse, and all the more so in less-regulated countries.[2]

While the environmental effects of endocrine-disrupting chemicals may mostly go unnoticed unless specifically looked for, other chemicals bleed off into the atmosphere in a way that can be instantly detectable. Practically every product in the cosmetic or 'personal hygiene' market is chemically scented. Even products claiming to be unscented will likely contain chemicals that mask any unpleasant smells given off by the other ingredients. How these molecules interact with each other or with plants and animals is almost entirely unknown. Men do often wear an ill-judged aftershave or deodorant, but it is most commonly women who walk around trailing that unmistakable cosmetic smell, where the various fragrances in their soap, skin cream, powder, lipstick, shampoo, hairspray and deodorant trail behind them in a discordant cloud. This is not a sexist observation, merely one by a person with an acute olfactory sense who has spent a lifetime noticing smells and wondering how on earth people can live with this hideous clash of aromas. To smell this in wafts from a neighbouring diner in a restaurant is to me quite as bad as stale tobacco smoke in clothing and temporarily deadens the tastebuds.

I have often wondered what happens to all the evaporants from drying paint, lighter fuel, stain remover, nail varnish, hand sanitisers, road markings and the myriad other volatile fractions that stream off so many substances and products, including those of the great (and many not-so-great) perfume houses. At last it seems that serious attention is being paid to their possible effects on health and the environment in general. A 'Pollutionwatch' article in the UK's *Guardian* newspaper noted that the fossil-fuel-derived chemicals that evaporate from printing inks, adhesives, coatings, cleaning agents and air fresheners now dominate the pollutants that form ozone in summer smogs and can even exceed the effects of emissions from traffic. Yet the chemicals that evaporate from personal care products turn out to be just as serious atmospheric pollutants. In an experiment volunteers were asked to shower using the same standard supermarket shower gel, shampoo, conditioner, moisturiser and finally aerosol deodorant.

Highly reactive limonene came mainly from the citrus-smelling shampoo, benzyl alcohol from the conditioner and ethanol from the moisturiser. This was different for each person and those who rinsed for longer produced fewer emissions. Other chemicals were seen, possibly linked to laundry products used to wash each volunteer's towel (they brought their own) or their clothes. It was also found that products worn by other researchers affected the air in the laboratory.[3]

In short, it is high time to bring pressure to bear on manufacturers by making them responsible for pollution by all these chemicals that go to make the collective stench of fake fruit flushed from a million shampoos in a zillion bathrooms, most of which will sooner or later reach a river system and finally the sea, with yet unquantified effects on the natural world. Above all, health-conscious consumers need to be educated away from equating cleanliness with perfume rather than with an absence of odour. The very idea of a deodorant that smells is an absurd contradiction in terms.

COMPLEMENTARY MEDICINES

The sheer size of the Wellness 'feelgood' market admittedly makes any radical change seem a hopeless ambition even though the natural environment that surrounds us is invisibly as well as visibly deteriorating. We should not overlook traditional and complementary medicine, a sector worth $360 billion on its own. In Britain, branches of Holland & Barrett ('The UK's leading health and well-being store') are a familiar sight in most towns of any size. The company self-describes as the largest Wellness retailer in Europe. There are thousands of similar shops for 'health and well-being' products in Britain and Europe and probably tens of thousands across the United States. There is a superb irony here since their customers are, for the most part, the healthiest, best-fed and best-medicated in the history of the human race. Yet here they are, spending fortunes on capsules and lotions and extracts of ever-more arcane plants.

Hardly a year goes by without some new 'ancient remedy' being discovered. A tale is spun about the astonishing curative and even mystic qualities of a weed that nobody had noticed before. Someone claims to have found Monk's Benison in a twelfth-century Latin manuscript describing the herb garden of a prioress in Aquitaine. The 'almost uncanny' powers of Monk's Benison are duly confirmed by Dr Ho Listik's research institute. He knows from personal experience that a single capsule of the extract can shrink your prostate in three weeks. Not only that, but it contains a substance, p-elixironic acid, that is known to boost intelligence and rejuvenate hair follicles to an extent that no one would believe until they try it for themselves... And suddenly the shelves of complementary medicine shops have apologetic notices lamenting that the run on the popular Monk's Benison capsules has regrettably left us temporarily out of stock but more are on the way and should reach us by mid-June. Word of mouth (aka word of social media) does wonders.

This industry, together with its myriad high street outlets, constitutes the Great Placebo Conspiracy. Millions of customers worldwide believe they detect improvement in their health, usually in direct proportion to how much they spend on things like Manuka honey following the discovery some years ago that it was actually the elixir of life. Manuka is now probably rather passé, having been superseded by Monk's Benison, Tibetan Dongweed and the rest. All this is fine: people must pamper themselves as their sense of humour permits. (One of the lines sold by Holland & Barrett is called 'Beautiful Me' and I dare anyone to present a jar at the cash desk without a blush.) What is definitely not fine is the environmental damage caused by things like fish oil.

FISH OIL

There may or may not be such things as renewable plantations of gingko and neem trees, but the stocks of tiny fish presently caught by the million tons for their oil are heavily threatened. Not only that, but depleting their numbers in the wild threatens starvation for other species that depend on them for food. This pattern goes back a long way, and includes the enormous hauls of Antarctic krill – a tiny shrimplike crustacean – by Soviet trawlers that decades ago caused starvation for whales, seals, squid, penguins and albatrosses. Then there was the sand-eel fishery of Western Europe, when not so long ago these little fish were hauled out by the million tons to be ground up for fish meal to feed livestock or ploughed straight into fields as fertiliser. In the Hebrides, puffin, kittiwake and auk communities began crashing until the sand-eel fishery was regulated. And now it is the turn of Peruvian anchovies, hauled out in pitiless quantities to be reduced to an oil containing omega-3 fatty acids. These substances are important for maintaining cellular health and are present in a wide range of foods. It is much the same as with hyaluronic acid. Almost everyone in developed countries has enough of both; but once the word 'essential' is tacked onto capsules containing omega-3 acids and they are heavily promoted as dietary supplements, there is no limiting the predatory fishing that goes on far over somebody else's horizon.

But that is not all. I once tried some of the oil in a teaspoon and knew immediately that it was rancid. It had that unmistakable oxidised taste that all oils and fats can acquire, from butter to olive oil, when they are exposed to the air and warmth. A recent newspaper article has thrown light on this. According to Richa Syal writing in the *Guardian* on 18 January 2022, each year an estimated 38,000 tonnes of anchovy oil are extracted from over 4 million tonnes of Peruvian anchovies for this global market worth billions. This tonnage of fish is way beyond the total annual catches of many a nation and represents an awful lot of live wild creatures to kill, not for food but on the outside chance of making healthy people still healthier. It is indeed strange that a Wellness market that nowadays is so imbued with vegan, vegetarian and environmentalist values should cheerfully connive at this wholesale slaughter of living creatures, and that the Holland & Barretts of the world should continue to promote the products of the trade.

It is hardly surprising that a high proportion of the oil should be rancid. Once the crude oil has been crushed out of these little fish in Peru it is shipped to China for refining and processing, which takes time and exposure to air and rising temperatures, both of which help oxidise what is essentially a delicate, cold-water oil. The distilled oil is then re-exported to Europe and the US for packaging. By the time it reaches the shops in capsule form or else as free oil to be taken (like cod liver oil) by the spoonful, the chances are excellent that it will be rancid.

ADVENT CALENDARS AND BEYOND

On 15 September 2021 British readers of the *Mirror* found themselves treated to a special offer that managed to combine complementary medicine with pampering. Hands up anyone with the nerve to take a stand against some copywriter's inspiration, 'clean beauty', by wondering what dirty beauty might look like, or even clean ugliness, come to that:

> Fans of clean beauty are in for a treat, as Holland & Barrett have launched their brand-new beauty advent calendar for 2021.
>
> The high street chain's all-natural beauty advent calendar, worth a whopping £170, went on sale this week and you can buy it from their website for just £45.
>
> This year's Christmas countdown is packed with a selection of pampering skincare and hair treats from brands like Dr Organic, REN, Weleda and Sukin. We can't think of a better gift to treat your favourite clean beauty fan to.

In Consumerland 'Advent' has lost any vestigial religious significance and like all sales gimmicks just means 'Stuff at bargain prices!' Even while many families are still on their summer holidays in early- to mid-September and with at least a quarter of the year yet to go before Christmas, the first 'Advent calendars' are already starting to appear. They herald the birth of a venerated economic saviour: maximising sales of a company's products by packaging a lot of them together at a reduced price and calling it a gift. This is the old 'two for the price of one' or BOGOF ploy updated to cover a bunch of products claimed to be priced at £170 but yours for a giveaway £45, which is still probably about £37 more than they actually cost to produce. The only thing *Mirror* readers needed to cudgel their brains over was who among their friends qualified as their 'favourite clean beauty fan'. That, and whether they thought they were worth presenting with £45's worth of pampering skincare and hair treats. Two months later in late November 2021, Black Friday Offers included several different 'Advent calendars' that would spill a new goodie each day in the run-up to Christmas: the greatest goodie-fest of them all.

That year Gwyneth Paltrow's Goop gift selection included well over a million dollars' worth of products starting with an 'eco-friendly floating hotel suite' for $535,000. This was the 'Athénea pod', a sort of floating igloo with windows above and below the waterline. It was not clear what it was for, but if you couldn't afford it you could lower your sights and make do with a gold and diamond Cartier watch-and-bracelet for $393,000. Thereafter it was downhill all the way with children's indoor playground equipment made of wood, velvet and gold plating at a mere $37,290, felicitously described by Goop as having 'rich Mom energy'; or a Hermès Birkin bag at $18,500. Positively slumming it were two coffee table books at $4,900 and $2,250 apiece until you got to the Wellness travel gifts. These included a trip to Rajasthan to see tigers; a stay in a Swedish treehouse hotel at $606 per night; a 'Home Alone 2' deal at the Plaza Hotel in Manhattan starting at $1,000 a night; and swimming with sharks near Cape Town for a measly $500+ for a weekend. Should you feel such daytime excitement might leave the nights a bit lacking in thrills, Goop recommended ten different vibrators including one for $180 'that makes a come-hither motion against the vaginal wall' and another for $129 from Smile Makers called 'The Poet', designed to create 'sucking sensations' (which, as any schoolchild will tell you, is just what poetry does best). However, anyone living in Texas would be advised not to collect all ten of Goop's vibrators because under Texas law they are ranked as 'obscene devices' and ownership of more than six is presumed to intend 'wholesale promotion' of them and breaks the law. One may, of course, legally own more than six guns in Texas, but not dildos.

Elsewhere, far away in the real world in December 2021, entire towns in Kentucky and Arkansas were reduced to matchwood by tornadoes probably whipped into ever more savage power by global warming. A couple of weeks later in his Christmas Day message Pope Francis, terrible old Cassandra that he is, warned that the world has now become so desensitised to crises and suffering that they are barely noticed when they happen. Goop would probably say that what he needs is a good night's earnest communion with 'The Poet' to get his mind straight and his body filled with the divine sensation of Wellness.

HEALTH TOURISM

Where health tourism is concerned, the Wellness variety and the Spa industry can be joined at the hip. Both stress the idea of physical cosseting. In May 2021 the ineffable Ms Paltrow again turned up in the news. 'I am always happiest by, in or on the sea!' she burbled to the *Miami Herald*. 'In 2022 my @goop team and I are going to join @celebritycruises on their new ship, *Celebrity Beyond*. I'll be behind the scenes, working on some special projects as Celebrity's new Well-being Advisor. My team @goop is curating programming and fitness kits to add to Celebrity's Wellness experience. I'm sworn to secrecy on the rest.'[4] The cruise line's CEO, Lisa Lutoff-Perlo, promised that guests would 'be gifted swag-bags when they board, chock-full of Goop'. The ship itself is described by its company as having 'innovative, outward-facing design leveraging endless ocean vistas and breathtaking panoramas,' which is American for saying you can see the sea from the ship. With luck guests will discover The Poet in their swag-bag to keep them company in the night hours and help them 'focus on nourishing mind, body and spirit'. An unexpected note of seriousness is struck by the news that Celebrity Cruises is one of the first companies to offer fully-vaccinated sailings where guests aged 18 and over must have had their Covid shots.

If so, *Celebrity Beyond* probably had a good chance of sailing from Barcelona to Rome as planned on 24 September 2022. There's Wellness but there's also Illness, and other cruise lines have been hard hit by outbreaks of Covid.

Over Christmas 2021 the Carnival Freedom was banned from landing in Bonaire and Aruba because the virus was spreading rapidly among the passengers aboard, to the acute disappointment of one lady who had always wanted to cross Bonaire off her bucket list. 'We're sailing on a petri dish,' she told a reporter. 'I feel like I just spent my past week at a superspreader event.'[5]

On the same December morning an unsolicited ad arrived in my inbox:

> Super spas. Our guide to hotels where you'll find sensational treatments.
>
> Relax and rejuvenate, lose weight and luxuriate, get primed and pampered ... and enjoy a wonderful holiday in the process.
>
> This week we're featuring 10 of our favourite Spa Hotels around the world where you can expect the very best treatments – and where there are tempting offers.
>
> Whether you fancy indulging yourself on an island in the Indian Ocean, chilling out in the Caribbean or de-stressing in Dubai, there is somewhere that fits the bill. So look for our great deals and flexible air fares, dust off those passports and check out these super spas.

Why do we even bother to point out that this sort of Wellness necessarily entails a heavy environmental price? The list of 'super spas' attached to the above email included their locations: the Maldives, St. Lucia, Seychelles, Thailand,

Mauritius, Dubai, Bali and Barbados. When fully costed in terms of aircraft, cruise liners, tour buses, hotels and even farting Egyptian camels, it is easy to see how those 830-odd million Wellness trips in 2018 added up to a massive carbon and methane footprint while accounting for 17 per cent of all tourism. This percentage alone was worth $639 billion, a figure that was quickly forecast to rise to $919 billion, although this may have proved optimistic as a result of the Covid pandemic.

But if, as with so many other sectors, Wellness for humans is striking yet another deathblow to the planet, it does also carry a related threat to its numberless practitioners with their shredded physiques and 'better-body goals'. Nowhere can one detect the faintest glimmer of interest in better-mind goals (and no, the blah of 'Mindfulness' does not count). It ought to be a perpetual source of wonder why the millions who anguish about their bodies aren't more worried about their pitiful, self-centred, consumerised thought processes. It is obviously so much easier to wolf down protein bars and jiggle to music in a leotard than it is to think. If it were not, the purveyors of outrageously foolish 'aids' to health for the already healthy would not find their task so easy.

SCAMS

In the fertile field of Wellness, scams are numerous beyond counting, but it is enough to take a single example. Himalayan salt lamps are typically chunks of pink rock salt with a lightbulb inside. They are sold as highly beneficial with the claim that when switched on they diffuse negative ions into the air, and negative ions make one feel good and even improve mental performance. Positive ions, on the other hand, allegedly make moods worse and so are a bad thing. (In order not to add to the fog of bafflement we should explain that an ion is an atom or a molecule that has gained or lost an electron. Losing an electron turns it into a positive ion and gaining one makes it a negative ion.) Negative ions are being constantly generated by a variety of natural events such as cosmic rays, lightning strikes, breaking waves and even from sources of radioactivity such as granite – meaning that the people of Edinburgh ought to feel exceptionally bobbish and be unusually quick-witted. Science has been investigating ions for at least a century, at least partly to discover why some people's moods appear to improve around waterfalls, on beaches or after thunderstorms.

In an episode on his often-impressive science site Veritasium, Derek Muller duly bought a Himalayan salt lamp and took it to a Caltech laboratory where, having warmed it up, an ionisation expert found it was producing precisely zero ions of any description. Asked for an explanation as to why he should expect any, Muller repeated the salesman's claim that water molecules landing on the salt's surface were supposed to liberate negative chloride ions. He was assured by the expert that this would require vastly more energy than the output of a simple salt lamp so he was unsurprised to find this one emitting no ions whatever. Obviously such lamps are a scam: not only do they not work, but they couldn't.

However, having noted that 'evil' positive ions tend to predominate in polluted areas around factories and in big cities, Muller concluded sensibly, 'I think ultimately if you are looking for a way to improve your mental and physical health that is backed by strong scientific evidence, then you should simply go for a walk.'

In terms of promoting health, simply going for a walk is probably the most beneficial measure for people and environment alike. But there's no money in it; no global Wellness industry, no pumping iron, no aromatherapy, no rowing machines and Bach flower remedies, no protein supplements and above all, no Gwyneth Paltrow. So it won't work.

Personal Freedom vs the Planet: Cryptocurrencies

A dozen or so years ago it all seemed innocent enough. When Bitcoins first hit the headlines the idea of a cryptocurrency seemed to recommend itself for all sorts of reasons. It was the first digital cash system and at the very least it offered the environmental advantages of being paper-free. No printed currency was involved, nor any of the mountain of statements and other material that banks regularly send through the post to their customers. Moreover, it offered a way of moving money around the world anonymously, making transactions without involving banks, local banking laws or the taxman. To many – and especially to certain US Republicans – it was seen as one of the welcome signs of a revolution enabled by technology that would at last strike a blow for freedom and independence. 'With the rise of Bitcoin and other cryptocurrencies we are fast approaching a time when central banks (and their client governments) no longer maintain a monopoly over currencies that allows politicians and bankers to tax, punish, spend, and control citizens' wealth and livelihoods from the shadows of elite and insular economic clubs', as a gentleman named J. B. Shurk wrote in the online site American Thinker.[1] Hillary Clinton soon made the Democrat case. At the Bloomberg New Economy Forum in Singapore she warned, 'What looks like a very interesting and somewhat exotic effort to literally mine new coins in order to trade with them has the potential for undermining currencies, undermining the role of the Dollar as the reserve currency, for destabilising nations, perhaps starting with small ones, but going much larger.'[2]

In most respects – and to most people – Bitcoin remained an enigma. It was apparently founded by 'Satoshi Nakamoto' – who may be an individual of any gender or nationality or possibly a group, no one seems to know. Since then, dozens of other cryptocurrencies have been launched (Litecoin and Ethereum, for example), and although a number of companies have accepted cryptos for commercial transactions, they are still not in very widespread use, in part because they are so volatile. In May 2021 Elon Musk stopped people buying his Tesla cars using Bitcoin, causing the cryptocurrency's immediate crash. Meanwhile, governments have been desperately trying to reconcile the anonymity of blockchain technology with their national accounting systems, most particularly for matters of taxation and crime.

Since cryptocurrencies are an intangible alternative to traditional money and users carry out financial operations such as payments using unbreakable algorithms, it is still next to impossible for governments to keep track of individuals' transactions. This has made them attractive for illegal purposes. Bitcoins were extensively used on the so-called 'dark' website Silk Road (until the site was shut down in 2014) for the anonymous purchase of highly regulated or outlawed things such as guns, drugs and child pornography. Since then it and other cryptocurrencies have been increasingly used for darknet purposes. At the beginning, the idea of a 'dark' web seemed almost science-fictional, or at least esoteric beyond the capabilities of most computer users other than the most dedicated nerds to grasp. It has now become easier to access via software like Tor and I2P and the anonymity of using VPNs (virtual private networks). Thanks to this more people are reaching the necessary degree of nerdiness. The police often appear to be fighting a losing battle, in particular with drug dealers. As soon as they shut down one site, two more spring up. The anonymity of the dealers, their customers and the cryptocurrency itself – all protected by encryption that is allegedly impossible to break – pretty much guarantees the hydra-headed burgeoning of these sorts of transaction.

Like all the other wizard new technologies we have ecstatically embraced before discovering the mischief in the package, cryptocurrencies pose the impossible problem of squaring anonymity with accountability. In the digital age, anonymity and black markets go hand in hand and offer a new layer of threat to assorted authorities. This may be excellent news to people who share US Republicans' views on the absolute sanctity of the individual over the state; but it is bound to be less so to those who accept that man is a social animal and social animals need to live together in as orderly a fashion as possible, otherwise everything quickly falls apart (as Americans have recently been discovering for themselves).

The advantages of cryptocurrencies are easily summarised. They offer virtual money with real convertible value. The money is held in an individual's 'wallet', a database known as 'the blockchain' accessed via an encrypted password. (In fact this is often how police manage to obtain a prosecution. In the manner of people everywhere, some write down a reminder of their password, often in an easily broken code and sometimes even en clair.) Transactions are quick, and unregulated except by their users. No change to the system itself is possible without the consent of every single customer. This is because the blockchain file is stored on a global network of computers and every transaction is readable by all users. The system is thus both transparent and impossible to alter. To that extent it is safe: each user has unique cryptographic keys for carrying out transactions, which go through immediately. E-commerce thus offers the opportunity for individuals and companies to make quick unmonitored payments. This is particularly useful when it involves countries that do not share banking treaties because it avoids having to go through 'normal channels' of officialdom that might otherwise take several days.

Among the various disadvantages of cryptocurrencies are that transactions are irreversible. Once paid, they cannot be recouped except by mutual agreement. If you send money to someone by mistake and they simply refuse to refund it, it's gone and you have no recourse. An additional drawback is the one usually touted as cryptos' outstanding feature. As one website claims, 'Since Bitcoin uses a complex algorithm, it cannot be manipulated by any individual, organisation or country, as some crazy serious skill would be required to carry out a digital heist.' To which one's instant retort is that if there is one thing the computer age has taught us, it is that digital security is constantly trying to stay one step ahead of talented hackers equally keen to break your 'unbreakable' algorithm. It is often said that tomorrow's quantum computing will be able to break all present-day encryption and so the endless pursuit for absolute security will go on. As we shall see, crypto scams are already quite possible.

Meanwhile, we have all been warned by graphic stories that cryptos can lose individuals huge fortunes, less by scams than by simple human carelessness. In 2013 James Howells accidentally threw away a hard drive containing his wallet loaded with 7,500 Bitcoins. It wound up in a Welsh rubbish dump. Journalists have variously assessed the coins' value at £200 million or perhaps $90 million, depending on the currency's fluctuations. In January 2021 the luckless Mr Howells offered Newport City Council £50 million to bulldoze through their heaps of refuse in search of his hard drive. The council refused, saying not only that it was not worth their employees' effort to look for something so small and long buried but also that the excavations would have 'huge environmental impact on the surrounding area'.

Still more on the drawback side, cryptocurrencies are very far from being a universal panacea. For a start, they are hard to understand and therefore to trust by anyone who is not computer-savvy. This extends to many companies and not just to individuals. More importantly, such currencies have proved highly volatile in value; and while investors might make a great deal of money very quickly if the value rises, they equally stand to lose if it falls. Early in 2021 a crypto called Dogecoin rose a hundred times faster than Bitcoin while a third currency, Ethereum's Ether, had recently hit a record high. In May 2021 Andrew Bailey, the Governor of the Bank of England, felt it necessary to point out that cryptocurrencies have no intrinsic value. 'That doesn't mean to say people don't put value on them, because they can have extrinsic value [...] Buy them only if you're prepared to lose all your money.' Indeed, the inherent instability of cryptocurrencies was dramatically exposed in early September 2021 even as Bitcoin was adopted as legal tender by El Salvador. An estimated $400 billion was wiped off the various cryptocurrencies' market value in a matter of hours. Four months earlier, as already mentioned, Elon Musk's ban on Tesla buyers using Bitcoin had caused it to fall heavily.

This extreme volatility is a good reason why many companies and individuals mistrust cryptos. Nor are such currencies accepted everywhere. They are already illegal in China, Taiwan, Vietnam and Bolivia and the list is constantly lengthening. On the other hand in early 2022 Gibraltar (of all places) announced that the crypto blockchain Valereum plc had bought Gibraltar's stock exchange, so might soon be trading conventional bonds alongside cryptos like Dogecoin and Bitcoin. This would put Britain in a somewhat equivocal position. Even as the UK warns strongly against investing in cryptocurrencies, the British Treasury is itself reported to be assessing the possibilities of a central digital currency tentatively named (what else?) Britcoin. The then-Chancellor Rishi Sunak proclaimed, 'Our vision is for a more open, Greener and more technologically advanced financial services sector.' It is quite likely that his long-term hidden agenda was to do away with coins and banknotes and have all the nation's transactions achieved electronically, whether a major company paying out its shareholders' dividends or a pensioner buying a packet of jelly babies. Doing away with cumbersome old cash, which people have been finding useful for some 3,000 years, will hopefully be resisted to the death by millions of citizens, particularly those older than about forty, whose collective dying-off might well take a further fifty years.

It can be assumed that governments everywhere – especially in the EU – are working hard to tame cryptocurrencies for the purposes of tax and customs control. In December 2021 the Austrian government announced that from 1 March 2022 cryptocurrency assets would be taxed at a rate of 27.5 per cent, irrespective of how long the assets have been held. These sorts of move have no doubt been prompted by the fact that the market capitalisation of crypto assets worldwide has grown tenfold since early 2020 to some $2.6 trillion, or about 1 per cent of global financial assets. In Britain well over 2 million people now each hold some £300 worth of crypto assets such as Bitcoin, Ethereum or Binance. This ought to surprise few who know Britain as a land of betting shops. People everywhere have always speculated in currency values, and the fact that cryptos are a more than ordinarily volatile field is unlikely to deter those with the gambling instinct. On the contrary, it is already fuelling an addiction.

GAMBLING

Cryptocurrencies constantly come and go. Newly created issues popularly known as 'shitcoins' have no function other than as gambling tokens. They are launched, efficiently boosted by advertising and influencers' chatter on social media, and are bought by speculators hoping to cash out before they crash: a process known as 'pump and dump'. In 2021 cryptocurrency companies launched a huge promotional push in London with ads on billboards, double-decker buses and at tube stations. This had a considerable effect, with dozens of start-up shitcoins attracting attention. Football clubs and individual players had already been touting their crypto investments daily on social media. The addiction expert Dr Anna Lembke, professor of psychiatry at Stanford University, commented, 'When you mix social media with financial platforms you make a new drug even more potent.' One of the catchphrases on social media is the acronym FOMO, or fear of missing out. This is widely used in connection with cryptos and helps ginger up the urge to buy into one or other shitcoin before it's too late. As Dr Lembke says, 'You get this herd mentality where people talk to each other constantly about what the market is doing. They have wins together, losses together: [it's] an intense shared emotional experience. We get a little spike in dopamine, followed by a little deficit that has us looking to recreate that state.'³ None of this is yet regulated in the UK. The Financial Conduct Authority (which *Private Eye* habitually mocks as the Fundamentally Complicit Authority) is powerless to act without new legislation. In addition, the government of the day was successively paralysed by Covid, an NHS in a state of near-breakdown, galloping inflation and an out-of-his-depth prime minister who was pungently described as 'A dead man walking' by one of his own senior Tory MPs.

GLOBAL ENVIRONMENTAL COST

It was Britain's former Chancellor Rishi Sunak's use of the word 'Greener' in relation to financial services that perhaps betrayed a radical misapprehension. Forgetting the probable social damage for the moment, cryptocurrencies come at a massive environmental cost even when not having to be re-mined from Welsh rubbish tips. In 2018 researchers at the University of New Mexico calculated that every dollar of Bitcoin value was responsible for forty-nine cents' worth of health and climate damage in the US. This is down to the prodigious amounts of electricity required by computers in data centres worldwide as they solve the complex algorithms to mine new blocks in the chain. In September 2020 *Smithsonian* magazine estimated that such data centres already use roughly 1 per cent of the world's electricity (by now that figure is considerably higher). In the US alone data centres consume as much electricity and water in a year as does a city the size of Philadelphia. The pollution associated with simply generating this much electricity is only part of the problem: still more power is needed to run the immense air conditioners and coolers needed to keep the temperature-sensitive computers functioning. Then on 29 September 2022 *Mailonline* quoted researchers as saying that Bitcoin mining uses more energy per year and generates more carbon emissions than the whole of Austria and is more environmentally costly than beef production or mining precious metals such as gold and copper. This was a trend that *New Scientist* had already predicted.[4] Rather than being considered akin to 'digital gold', Bitcoin should be compared to energy-intensive products such as natural gas and crude oil.

In terms of electrical energy, completing a single Bitcoin transaction requires around 1,500 kWh. This is roughly fifty-three days' worth of power for an average household. As of writing, a kilowatt hour in Britain has been capped at fifty-two pence, so 1,500 kWh is equivalent to some £780. Cryptos naturally choose to mine their currency where electricity is cheapest. This is why, until China banned it, 75 per cent of all mining took place there. The ban was accounted a good thing from the environmental point of view since most of the electricity needed had been generated by coal-fired power stations. However, much of the world's crypto mining promptly switched to the place with the next-cheapest electricity: Kazakhstan, which quickly became the second-biggest centre of Bitcoin mining after the US itself. According to *Fortune*, by August 2021 Kazakhstan was hosting 18 per cent of all the world's cryptocurrency mining. Environmentally speaking, this was worse news because Kazakhstan's power generation relies heavily on ancient Soviet-era coal-fired power stations that chuck out even more carbon than do China's: half as much again per kilowatt hour, according to the International Energy Agency. Then in January 2022 Kazakhstan experienced violent riots during which Internet access was shut down. This immediately brought Bitcoin mining there to a halt and its value once again sank. For the planet as a whole this was no bad thing; but in the long run it probably made little difference, given cryptocurrencies' increasing popularity.

In 2021 a US crypto company, Marathon, bought and took over a moribund coal-fired generating station in Hardin, Montana, solely for its 115-megawatt capability. On an adjacent twenty acres it built a data centre and filled it with over 30,000 Antminer S19 computers specialised for mining cryptocurrency. By the third quarter of the year the resurrected plant was emitting 206,000 tons of CO_2... As Anne Hedges, a co-director of the Montana Environmental Information Center, bitterly observed, 'Re-starting the power plant isn't preventing old ladies from freezing to death. It's to enrich a few people while destroying our climate for all of us. If you're concerned about climate change you should have nothing to do with cryptocurrency. It's a disaster for the climate.'[5]

Some people did read Elon Musk's refusal to accept Bitcoin as a hopeful sign. Maybe people with influence would take the environmental consequences of cryptocurrencies to heart.

The reason Musk gave was that he was 'concerned about the rapidly increasing use of fossil fuels for Bitcoin mining and transactions, especially coal, which has the worst emissions of any fuel'. He added that, 'Cryptocurrency is a good idea [...] but this cannot come at great cost to the environment.'[6] Not everyone was won over by his apparent display of a Green conscience, however. As Julia Lee of Burman Invest observed, 'A cynic might suggest that this is just another move by Elon Musk to influence the cryptocurrency market, as he has done on so many other occasions.' Rory Cellan-Jones, the BBC's technology correspondent, noted that in view of Elon Musk's often-repeated concerns about the environment that expressly led to the development of his Tesla cars, it was somewhat surprising that he had only just woken up to the fact that Bitcoin is not exactly a Green project. Cambridge University's Centre for Alternative Finance runs a Bitcoin Electricity Consumption Index. The same university's Judge Business School claims that Bitcoin alone requires 144 terawatt-hours per year: more electricity than Argentina or Sweden.

Crypto enthusiasts may insist that mining mainly involves renewable energy. This is nonsense. Miners focus on whatever is cheapest. When he announced his grand cryptocurrency plans back in February 2021 – and made Tesla's chief financial officer 'Master of Coin', a wonderfully medieval title – Elon Musk sparked a surge in the value of Bitcoin and the adoration of its fanatical devotees. When the price fell again the chatrooms were full of the angry and disappointed, accusing their former hero of betrayal.

A move which was hailed as signalling that Bitcoin was going mainstream has ended up showing just how volatile – and downright flaky – the whole cryptocurrency scene can be. Mr Musk has been one of the world's highest-profile advocates of cryptocurrencies, often tweeting about Bitcoin and the once-obscure digital currency Dogecoin. His tweets in 2021 succeeded in turning Dogecoin, which was started as a social media joke, into the world's fourth-biggest cryptocurrency.[7]

NFTS AND ASICS

It should be noted that cryptocurrencies are not the only users of blockchains. The same technology is used to store the data of non-fungible tokens (NFTs). Like so much else in the digital world, these strike the uninitiated as both incomprehensible and daft. Yet somehow they generate oodles of hard cash. NFTs are virtual 'objects' with a blockchain-based proof of ownership and, like cryptos themselves, come under the broad heading of 'Decentralised Finance' (DeFi). NFTs are usually digital 'artworks' that are created, bought and sold. The digital artist named Beeple created a collage, *Everyday: The First 5000 Days*, that was auctioned by Christie's in March 2021 for $69.3 million. The actual artwork is merely a mass of billions of electronic pixels but the dollars are real. Not surprisingly, DeFi has become a lively field and other ingenious uses of blockchains – including virtual gaming – are constantly being dreamed up, adding still further to the environmental burden. 'In December 2020 the artist Memo Akten estimated that the average carbon footprint of a single NFT artwork was equivalent to driving a conventionally-engined car six hundred miles.'[8]

Even more recently news has come of another aspect of Bitcoin mining that is having an increasing effect on both the environment and the market for everyday computers. In order to 'mine' cryptocurrencies at a profit, new and ever more powerful computer chips are in constant demand to run the algorithms. These are ASICs (application-specific integrated circuits); and such is their speed of development they quickly get out of date. 'The lifespan of Bitcoin mining devices remains limited to just fifteen months or so. As a result the whole Bitcoin network currently cycles through 30,700 tons of electronic equipment each year, comparable to the amount of waste IT and telecoms equipment produced by a country like the Netherlands.'[9] The same source warned that the present demand for mining hardware like ASICs (which are not useful for anything else) could soon disrupt the global supply chains of semiconductors. This is already happening, with the chips needed for cars' computerised engines becoming so scarce that the second-hand car market is currently booming.

HACKING AND SCAMS

Despite the widely held belief that at least until quantum computing becomes common blockchain security is unbreakable, this does not mean that the system itself cannot be hacked. It was soon discovered that when a malicious group of a cryptocurrency's 'miners' (i.e., its software users) owns more than half a network's computing power, it can initiate a '51 per cent' attack. In 2019 a group of scammers with access to Ethereum Classic's blockchain employed a 51 per cent attack to rewrite its transactions history. In this way they stole over a million dollars' worth of the crypto from its other users. According to the blockchain data company Chainalysis, scammers stole $14 billions' worth of cryptos in 2021: nearly twice what was stolen the previous year. It never takes long for genius-level malevolence to find the flaws in any system trumpeted as 'unbreakably secure'. No doubt quantum computing, too, will turn out to be more permeable than its guileless boffins have claimed. So-called 'Ransomware' has already weaponised encryption in order to take possession of individuals' and companies' computer systems. The victims are generally given a day or two to pay a normally huge ransom in Bitcoin, on pain of having their precious customer and operating files irretrievably trashed. Bloody digital warfare rages daily throughout the world but for obvious reasons its victims hide their wounds in silence.

ANONYMITY VS CITIZENSHIP

In May 2022 cryptos suffered large losses worldwide triggered by the crash of Terra, one of the world's 'stablecoins' supposedly pegged one-for-one to the US dollar. When Terra was trading at twenty-three cents the crypto world panicked and even Bitcoin, the largest and oldest currency, fell to its lowest level in three years. By mid-2022 Terra was trading at a penny on the dollar. This sort of volatility, plus the scams and crashes and ludicrous NFTs, can give the impression that once all the hype has cleared away, cryptos (and Decentralised Finance itself) are probably doomed in the long run to fail.

Yet there are commentators who believe the scandals are actually doing DeFi a service by weeding out the fly-by-night speculators, the gamblers and chancers and darknet criminals. Some financial titans and institutions such as Morgan Stanley, Microsoft and Facebook have been quietly repositioning crypto technology as a powerful neoliberal tool. In an essay for *UnHerd* on 10 June 2022 David Auerbach wrote:

> The cryptocurrency crash is actually a blessing in disguise, thinning the ranks of small and sometimes disreputable competitors and investors while leaving the underlying infrastructure in place to be increasingly taken up by larger, institutionalised players. The wear and tear on the financial system is significant, but those celebrating the bursting of the bubble haven't yet noticed that a bigger game is afoot.

Yes indeed. In reality, crypto markets worldwide are in the hands of a few hundred people. Here we have the perfect example of how a new technology working to the benefit of a small, tech-savvy minority of the planet's inhabitants can not merely cause gross environmental damage to everybody's detriment. It can also generate fortunes great enough to stampede governments into coming to terms with entrepreneurs so wealthy that even heads of state fawn on them and are scared to tax their empires, squirrelled away as those are between tax havens scattered around the globe. Perhaps Hillary Clinton had a point after all. Yet even this manages to conceal a separate, but equally dangerous, downside. In many ways the digital age plays right into the hands of a growing number of people who see themselves as free citizens of the world, unaffiliated to authority of any kind. This is an increasingly common attitude among right-wing Americans (which by the standards of a majority of Europeans is most of them). Even those who are not financially secure are often fundamentally opposed to the very idea of government. Republicans in particular are prepared to defend their position with firearms, quotations from the Constitution and Trumpian demands to 'Drain the [Washington D.C.] Swamp!' The wealthier will devise ever more ingenious techniques of evading taxation and any other official accountability. The possibilities offered by virtual currencies and 'metaversal' identities make it increasingly difficult for hard-pressed officialdom to keep track of citizens and their activities. Doing so involves rarefied technology and suitably trained people and is enormously expensive – costs that are inevitably passed back to the rest of the population via taxes.

Such 'virtual citizens' are more than happy to avail themselves of all the services and advantages accruing to those with the sort of passport that guarantees them hassle-free international travel and an enviably wide range of civil rights at home, but they will ingeniously wriggle out of paying the taxes that underwrite those advantages. Moreover, they will do so without a shred of guilt, even if they are caught and made to pay, on the grounds that they see any governance as tyranny and refuse to recognise it as 'legitimate'. Many view themselves as having long since risen above the status of mere citizens to become free, independent, near-immortal inhabitants of the 'metaverse'. This apparently gives them the right to unlimited civil disobedience. Their very existence is empowered solely by the digitised world. It is a trend leading to the steady weakening of governments and their related agencies that alone can work internationally to prevent yet further degradation of such legal restraints as remain. Ultimately, social cohesion will thin and fray to everybody's loss; but for the moment the main casualty is the planet's health and this is (perhaps mistakenly) seen as more urgent.

To take a sad example of the decay the digital world can wreak on institutions, Britain's banking sector used to be a relatively upright system of considerable international renown. In the City of London of my own youth a banker's word was famously his bond and all transactions, though subject to privacy laws, were still in theory and practice open to official scrutiny. In much less than half my lifetime this once-workaday system has been – not allowed but *encouraged* – to degenerate to a point where London, like Britain's Overseas Territories in the Caribbean, has become a byword for international money laundering. Prodigious sums generated around the world by the drugs trade, prostitution, child pornography, people smuggling and the like pass every minute through London-branched foreign banks that add their own cryptic levels of anonymising code to the transactions so it becomes nearly impossible for any hard-pressed authority to keep a check on them.

Lord Faulks, a former Conservative minister in Theresa May's government, tried in 2017 and 2018 to put some teeth into two bills on criminals' finances and money laundering. Describing the anonymous ownership of dirty money in London as 'an obscenity', he wanted a public register of overseas property owners. On both occasions he was thwarted.

> Quite frankly, I was leant on. I was leant on by Number 10 Downing Street and told that 'the matter was in hand'. I was obviously misled because subsequently nothing has happened. I can only think a deluded desire to protect the City of London has led to all these delays. Over a period of five years successive ministers have offered me emollient promises that the issue was in hand. What has happened? Absolutely nothing. In the meantime, frankly, we look like a laughing stock. We are not responding to the threat of economic crime. Instead, we are giving away Golden Visas and the rest of the world must think we simply do not care.[10]

However, it is indeed an ill wind that blows nobody any good. It took the war in Ukraine finally to embarrass a British government into taking action. On 17 February 2022 the *Daily Mail* had this headlined account: 'Golden Visas that let anyone into the UK if they invest £2m will be scrapped to fight Russian "dirty money" as part of crackdown over Ukraine [...] A visa scheme that allows rich foreigners to live in Britain in exchange for a minimum £2 million investment is to be scrapped. There has been long-standing unease that the Home Office's "Tier 1" investor scheme was being abused by criminal gangs and the super-rich from overseas.' The Golden Visa system was introduced by Gordon Brown's Labour government in 2009 in an effort to attract investment to the UK. Since then more than 12,000 visas have been doled out, including 2,500 to Russians. Who could possibly have foreseen how this might go wrong?

Anyone who instinctively believes in some sort of democratic ideal for running a society would agree that in banking, as in all dealings between citizens and officialdom, the strictest privacy should indeed be observed. But that does not mean total anonymity. Sooner or later this privacy, given human nature, will be abused for nefarious purposes. Thus the steady increase of various law-flouting websites on the 'dark web' ought to have come as no surprise to anyone. At the first sign of policing they simply scurry further down the murky corridors of that vast, many-layered basement of the World Wide Web where most people lack either the computing skill or the inclination to venture. Complete anonymity is never wholly innocent.

Pessimists should take heart from what David A. Banks wrote in the *Guardian* on 12 May 2022. 'Crypto fans need to recognise a hard truth: that currencies and contracts are only as valuable or enforceable as the people and institutions that recognise their legitimacy. Blockchain technology does not change this fact.' In short, all financial markets need to be regulated. Unregulated but excitingly new ventures oversold by crafty as well as naïve news coverage sooner or later turn to dust. Their essence as a form of gambling using other people's money reveals them as Ponzi schemes whose investors, sooner or later, risk losing everything.

Shipping & Shopping

The most revealing – even depressing – result of calls for urgent action on behalf of the planet's health is that all the grand plans are primarily aimed at making our existing way of life less damaging; almost none suggest changing it. This is not surprising since the things that have most changed human behaviour over the last two centuries or so have never been moral arguments or the threat of consequences, and least of all have they been appeals to intelligence. Instead, they have been the technologies we invented: steam trains, telephones, cars, atom bombs, computers, Twitter and Facebook. Homo sapiens is not good at thinking far ahead. In well over 200,000 years of monkey-living, of hunter-gathering and hand-to-mouth existence, nobody needed to think much further ahead than possibly laying in some stores against the coming winter. So it is hardly surprising if none of the long-term consequences of any of our inventions have ever been seriously thought out in advance of their mass adoption. In reply to anyone advising caution we always wave an airy hand and say, 'Oh, there'll probably be some teething problems,' and adding, 'Anyway, we'll soon invent something to fix them.'

Nor are we much good at taking concerted action to mitigate the consequences of a technology once it has taken hold. Both cars and trains supply a good example. Why after the Second World War did we Britons not stop and think before allowing the vital matter of renewing the country's shattered transport infrastructure to be dictated entirely by the interests of the road lobby and private vehicle manufacturers? Why did we allow successive governments to dismantle large parts of the Victorians' brilliant rail network, abolish pollution-free electric trams and trolleybuses and cut back public transport in all but the biggest cities while leaving rural communities often dangerously isolated? Why have we spinelessly agreed to an incredibly inefficient and wasteful way of life that has locked us into having to make constant journeys for which we mostly need individual motor-driven transport on ever-proliferating roads?

Pointless rhetoric... In the last seventy years we have colluded almost unquestioningly with policies that have favoured powerful corporations and lobbies, smothered the landscape in tarmac and concrete, destroyed vast swathes of countryside while polluting it with carbon, lead and rubber particulates and worse. Once industries stopped being local and were swallowed up by great centralised corporations ancient patterns of work, employment and even shopping were changed for good. In what is after all quite a small country, we have so arranged economic matters and the logistics of living that we need to be constantly scurrying about like frenetic ants. Our lives drain away in commuting and shopping; we drive hither and yon at considerable expense, expelling and then breathing all sorts of atmospheric damage.

In retrospect this was planning failure on a grandiose scale, and what has happened at national level has equally happened at the global. The wealthier nations have created a world economy that now depends absolutely on long-distance maritime transportation by container ships, tankers and cargo vessels to the extent that these now handle well over 90 per cent of all international trade. Vast fleets are in constant motion, criss-crossing the oceans as fast as they can while leaving heavy trails of pollution. How does this implicate us, the consumer? The answer is up to our necks, because we have now reached the point where nearly everything we buy or use has spent some time in a ship, even if only as a component rather than as a finished article. Only in an outdated economic model could this make sense; yet all technical efforts now seem directed at making ships less damaging to the atmosphere instead of devising a way of life that could do without most of them.

(The Stuck Monkey stares in puzzlement at his own fist in the glass jar, but all he can really see is that banana. He yearns for it with a longing that dulls his intelligence.)

It might seem churlish to stigmatise our insatiable desire for *stuff* as the major hindrance to our planet's recovery, given that some variety of shopping is many people's major occupation. True, a great many heartily loathe this near-daily ritual, the dutiful and time-wasting expeditions to buy food and household necessities, and would be only too happy to stop having to do it. On the other hand there must also be very few consumers who don't thoroughly enjoy buying something for themselves (this would hardly be a revelation to Amazon). Willy-nilly we have made this a retail world and the entire global economy depends on it. The planet's future is inextricably bound up in it and it is now almost certainly too late to free it.

Back in March 2021 the huge container ship *Ever Given* got itself wedged across the Suez Canal for a week, blocking it and holding up some 370 ships forced to queue at either end. Each day's delay reportedly cost an incredible $9.6 billion. It showed how easily disrupted the global system can be. Even in late 2022 there are still serious international hold-ups in deliveries of goods to shops and other retail outlets, and container ships are queuing to get into ports worldwide, particularly in China and the US but also Rotterdam. This is partly because a shortage of truck and goods train drivers in the US, Europe and the UK has made it harder to move goods out of ports once they have been landed: a jam that port closures from the Covid pandemic only made worse. Shipping charges have shot up as a result. In early 2022 it cost £12,650 ($17,000) to send a single forty-foot container from Asia to Europe. A year earlier it cost a mere tenth of that.[1] The war in Ukraine and the embargo on Russian oil and gas has already led to a steep rise in international fuel prices, with an impact on retail prices worldwide. Even so, shipping was – and still is – unbelievably cheap. Shipping a single pair of trainers from China to California adds maybe one cent to their basic cost although recently that might have doubled or even trebled. The *Ever Given*'s expensive week in the Suez Canal reminded us that it is cheaper for Scottish fishermen to send their catches to China to be filleted and returned than it would be if the fish were filleted in Scotland: a perfect illustration of how the economics of our lifestyle have arranged for market forces everywhere to gang up on the global environment.

This sheer cheapness of mass shipping has literally moulded the world economy into a historically new shape because for most goods there is no longer any advantage in siting manufacture near your customers.[2] It is all part of what looks almost like a global conspiracy to promote the unceasing mechanised movement of goods and citizens, even to abolish stasis in any form. The blocked Suez Canal was the ideal metaphor for this: standing still costs a fortune each day and gets nobody anywhere, while being in constant motion brings home the bacon and everything else. Voltaire's 'Il faut cultiver notre jardin' suddenly sounds less like a commonsensical precept for a contented life than an old-fogeyish rejection of all restless modernity.

In order to circumvent the more general cargo holdups, several of the biggest retail outlets such as Ikea, Home Depot and Walmart resorted to buying their own containers and chartering vessels to carry them. This might in theory enable such companies to keep their prices down but it is still hard to see how it could get around the problem of pandemic-affected ports jammed with a backlog of ships. Huge container vessels can't just put into any old harbour and hope to unload. They need deepwater ports with giant cranes to deal with them, as well as acres of holding area for the offloaded containers to be piled in neat multicoloured ranks in the right bays for collection by road or rail. This puts a limit on the ships' potential destinations.

Any hiatus in shipping vividly reveals how dependent world trade is on the efficient containerisation of goods. The biggest ships can now carry 24,000 containers, each twenty feet long. Such vessels can take anywhere between six months and two or three years to build and might cost between $100 million and $200 million apiece: a huge gamble when the worldwide system can't be relied on to function smoothly. And over everything hangs the environmental question.

FUEL AND POLLUTION

For years now the world's shipping (which very much includes cruise liners) has been rightly tagged as one of the planet's most egregious sources of atmospheric pollution. Every year it alone has been responsible for putting some 940 million tons of CO_2 into the atmosphere: some 3 per cent of all global CO_2 emissions. Overall, this makes the sector a bigger contributor to environmental damage than aviation. The largest container ships together emit roughly as much pollution as 50 million petrol- or diesel-engined cars. The world's fifteen biggest vessels together emit as much nitrogen oxide and sulphur oxide as all the planet's 760 million cars.[3] The enormous engines of most cargo vessels still run on bunker oil: a thick, black, comparatively undistilled petroleum that needs to be heated to 50°C to overcome its viscosity before it is usable in an engine. The biggest ships can burn 300 tons of this stuff in a day. After enormous international pressure was applied to the industry over decades – not least by ports like Venice that regularly play host to cruise liners – the UN's International Maritime Organization (IMO) finally got its act together and belatedly imposed a limited system of Sulphur Emission Control Areas (SECAs). This regulation stipulates that any fuel used in these areas must have a maximum sulphur content of 0.1 per cent. Current SECAs include the North Sea, the Baltic and the English Channel as well as the entire 200-mile EEZ around the US coastline and Hawaii. In Europe this regulation is monitored by the European Maritime Safety Agency, carried out by patrols of 'sniffer' drones.[4] As from 1 January 2020 any bunker fuel burned in the rest of the world's oceans must have an absolute maximum sulphur content of 0.5 per cent: much less easy to monitor in view of the huge areas. More recently, the IMO has committed itself to a modest 50 per cent reduction of annual shipping emissions by 2050, perhaps recognising forecasts that by then the industry will have grown by anywhere between 50 and 250 per cent. Two and a half times the present volume of shipping will easily negate most if not all the advantage gained by more efficient engines and the not-very-radical curbs on exhaust so recently introduced. We are firmly in Jevons Paradox territory here (see page 83 of the Cars & Aircraft chapter). If the IMO is serious it will have to impose far more severe and globally enforced restrictions on emissions. The very uncertainty of such predictions, combined with the recent revelation of how easily political and other unforeseen circumstances can combine to bring the entire maritime transport industry to a global near-standstill, shows how difficult it is to envisage its future in the sort of hard detail required for long-term investment in vessels taking years and costing hundreds of millions of dollars to build.

Despite these belated signs of environmental awareness in the shipping industry, Britain's own position is likely to have been made worse by Brexit, according to an analysis by the UK Trade and Business Commission that was published – with maximum potential embarrassment for the host nation – just before the COP26 climate summit that opened in Glasgow on the last day of October 2021. The report made the sensible point that Britain's post-Brexit shift in trade links away from nearby Europe to far-flung partners such as China, Australia and the US would almost double the UK's greenhouse gas emissions from shipping, increasing them by 88 per cent. It was yet another unforeseen aspect of what had already begun to look like an economically disastrous political gambit championed by Prime Minister Boris Johnson himself, the very man hosting the climate conference. 'The release of an additional 6.5 million tonnes of carbon dioxide into the atmosphere each year by container ships making long voyages around the globe with goods for the UK market and exports to overseas customers would be the equivalent of 44,000 transatlantic flights, the Commission found.'[5]

There have recently been signs of a change of heart among the owners of these fleets of giant vessels by commissioning new ones that run on methanol. There are already a couple of dozen cargo ships powered by this much cleaner fuel but the breakthrough came in August 2021 when the world's biggest shipping company, Maersk, announced it had ordered eight new giant container vessels that could be run on both low-sulphur oil and methanol. This is better news from an environmental point of view, although two major factors act to discourage widespread adoption of methanol as a fuel. The first is cost: methanol is about twice as expensive as oil. The second is its comparative scarcity. World methanol production currently runs at some 200,000 tons annually and Maersk reckons that its eight new ships will together need twice that each year. At the moment methanol is produced commercially in two ways: either from biomass or as e-methanol synthesised by ingenious industrial chemistry out of Green hydrogen and CO_2 that has been removed from the atmosphere (see Appendix 2: Biofuels for details). Currently, the biomass method is cheaper but has environmental limitations. It will cost time and money to step up methanol production no matter which method is used, and Maersk and other lines will of course pass the costs on to exporters and importers in increased shipping fees. This will duly raise retail prices in shops, although margins are already so low it may not make much difference to consumers. In the meantime Maersk will probably be undercut by German and French shipping lines that are opting for liquid natural gas. This is much cheaper and takes advantage of some regulations that favour it. This despite frequent methane leaks from wells, pumping stations and the ships' engines themselves that mean

LNG can itself have a deleterious impact on the climate even before it is burned in an engine: one of the reasons Maersk decided to avoid it.

Another quiet revolution in shipbuilding is a general trend towards electric propulsion. Ships, and particularly maritime research vessels, have long adopted such things as thrusters: subsidiary propellers below the waterline on either side of the hull that make it possible to manoeuvre accurately and even to hold a vessel stationary above a point on the sea bed in water too deep for an anchor. This has led to the notion of driving ships by means of several power pods instead of by one or more gigantic screws at the vessel's stern. Each pod has its own electric motor driving a propeller and can be fully swivelled. This means the ship is no longer dependent on going either ahead or astern and is much easier to stop or turn and fully capable of moving sideways: a huge asset when negotiating a crowded port. Podded propellers can also lead to a much less cluttered engine room. One problem with the historic set-up of engine and conventional propulsion is that it takes up a lot of space with its massive gearing and propeller shafts. The new technology does away with that by having generators that simply distribute electric current to the propulsion pods under the hull. This idea of fundamentally electrifying ships bestows much more flexibility. The generators can run on such fuels as methanol or ammonia or else in time could be replaced entirely by fuel cells. Where cargo vessels are concerned, battery power is not seriously considered at the moment. As Lee Kindberg, Maersk North America's Head of Environment and Sustainability remarked, 'If we actually wanted to propel a container ship just with batteries, we'd have no room left for cargo.'[6] It is conceivable that this verdict might change if ever batteries become substantially smaller.

SUPPLEMENTARY POWER

One possibility for an additional source of renewable power is by installing Flettner rotors up on deck. These were a clever idea by a German aviation engineer in the 1920s. He proposed using long end-stopped cylinders in place of an aircraft's wings. These were driven to rotate and produce the Magnus effect, whereby an airstream hitting the rotors from in front would be deflected perpendicularly downwards to provide lift. This was tried and found to work; but the system's obvious disadvantage was that if the engine spinning the cylinders broke down the Magnus effect would also cease and, lacking conventional wings enabling it to glide, the luckless aircraft would fall like a stone. However, when Flettner rotors were installed vertically on a ship they were indeed found to provide a degree of forward propulsion when the wind came from the side. There are now several ships with these thirty-metre-high rotors installed. Apparently they are rather noisy and of course no fitful wind power can take the place of engines. However, they can reduce fuel consumption by taking mechanical advantage of a side wind without the cumbersome and old-fashioned array of sails, masts and rigging that might otherwise do the same job. Some fifteen years ago the German company SkySails created a stir with their huge rectangular canopies that looked like giant versions of powered parachutes. These were deployed far above the ship to catch overhead winds and add their towing power to that of the ship's engine. The claim was they could save up to ten tons of fuel oil a day which, set against 300 tons, is not much; but where the environment is concerned every little helps.

There is one type of propulsion for ships that is by far the Greenest and most economical option: nuclear power. Its obvious advantages must mean that its lack of widespread adoption can only be attributed to a near-superstitious fear. The World Nuclear Association claims there are at this moment more than 160 ships worldwide powered by over 200 small reactors. Nuclear-powered submarines and other vessels – not all of those military – have been successfully in service for decades now. It is true that any nuclear reactor poses the same ultimate problem of decommissioning at the end of its life, but until then it can deliver years of carbon-free power with great efficiency. If the issue of decommissioning were finally and internationally addressed it would surely not be beyond human ingenuity to come up with a solution. There are many suitable places in the deep ocean where old reactors could be harmlessly entombed beneath a sea bed that has been geologically stable for millions of years. No doubt the very notion will raise an outcry among environmentalists; yet it would be by far the safest (if very expensive) solution to a problem that is already intractable on land. It would surely be preferable to the ongoing damage inflicted globally by the combustion of liquid fuels.

SLOW STEAMING

One expedient guaranteed to help reduce container vessels' pollution (but which is none too popular) is slow steaming. Cargo lines adopted this during the 2008 financial crisis when shipowners urgently needed to reduce their fuel bills. They did so by cutting their vessels' speed. CO_2 emissions – as well as those of sulphur and NOx – were thereby much reduced, reportedly by nearly 200 million tons: an amount roughly comparable to the annual emissions of a country the size of the Netherlands. If the IMO were to impose a reduced speed limit (these days easily monitored by satellite) the world's fleet of massive container vessels could very quickly become less damaging.

Of course longer delivery times would be unpopular with customers. Behind all this seaborne trade is the world's insatiable appetite that demands ever-quicker gratification from goods in quantities and varieties so numerous they defeat listing. Day and night, they are in constant motion by sea, by air, by land and via electronic downloads. If their paths over a year were physically plotted with threads on a large schoolroom globe it would soon be completely hidden beneath the strands of its own freight to fill the entire classroom with a gigantic ball of wool. These tangled windings represent all the supply chains and logistical requirements of industries and customers worldwide. It is one of those things that most people never think about, any more than they think about their car's engine. Our familiarity with modern retailing takes the constant availability of anything we might want for granted, and few of us have much awareness or even interest in what supports it. We have no real grasp of quite how vast the system is or of where things are made and the journeys they make to get into shops. They're just there. If ever there is a temporary hold-up in supply we moan about it, then we start to worry, and before long a single rumour spread on social media quickly leads to panic buying.

THE GLOBAL BAZAAR

A friend who lives in a small village in South-East Asia has often told me of a trip he once made to the United States – the only time he has ever been further than forty miles from home, let alone abroad. An uncle and aunt who had emigrated paid the airfare for him to spend a month with them in Los Angeles. He had barely arrived when they took him shopping to the local Kmart or Aldi or Walmart: one of those cathedrals of surfeit. His description of his first impression has never changed. It is of someone reduced almost to being a child again, overcome by the wonder of something unimaginable, beyond words, inconceivably majestic in its sheer, glittering abundance. At first he wanted everything he saw, imagining returning home to his palm-thatched village with armloads, *truckloads* of gifts, cornucopias of goodies. He knew what would most delight all his multitudinous family and friends: clothes, food, washing machines, steam irons, guns, computers, tools, rice cookers, glassware... And here it was, all under one roof, and it just went on and on in heaps, piles, mountains. My friend was well used to killing a chicken for a special occasion but was now dumbfounded to see packs the size of pillows full of frozen chicken wings or chicken legs, chicken thighs, chicken breasts. Dismantled animals of every kind filled gigantic freezer chests: enough to feed whole villages – maybe even a province. He was much taken by transparent plastic lavatory seats in which were embedded glittering fake aquarium fish in brilliant colours that made him laugh and yearn to buy some before remembering that back home he had a cement squat toilet in an earth-floored palm-frond hut. The following days in LA just added depth to his new awareness of the smallness of the world he himself lived in, where soy sauce was bought in little sachets and cigarettes in ones and twos. Suddenly his home village shrank to insignificance and he was gradually overcome by an eroding sadness of disparity. His uncle promised he would help with the whole rigmarole of sponsorship, Green Cards, immigration fees and all the rest of it if my friend wanted to join them in LA; but he never has. His sadness unaccountably turned into homesickness and he says he was glad to leave LA: one of the few people I have ever known turn down the offer of a trip on that magic carpet to the promised land of abundance far over the horizon.

Although whatever childhood wonder has long worn off, even accustomed shoppers do sometimes become aware of just how these vast supermarkets can affect their own behaviour. In October 2021 the *Daily Mail* published a revealing article about an American teacher in Sydney who told her Australian husband she was just nipping out to the local Kmart for a pair of socks. Describing her experience later as like 'falling down a rabbit hole' she emerged clutching candles and T-shirts and Tupperware pots, arriving home with everything except the socks she had gone out to buy. Various other women confirmed her shopping experience, one adding: 'The Special Buys aisle in Aldi is even worse: you end up with an arc welder or something else you never needed.'[7] Such retailing psychology is, of course, central to the global economy. The sheer allure of things you didn't know you wanted – still less needed – criss-crosses its daily trails of smoke beneath Pacific and Atlantic skies, much as the airlines' contrails far overhead fray and smudge in the upper atmosphere.

Finally, one important aspect of shipping worth noting is that the word itself has slipped – mainly through American usage – to include all the ways in which articles may reach the consumer, particularly via deliveries by road. When Amazon announced same-day shipping for Prime members, other companies like Walmart were obliged to follow by offering faster delivery. According to one research firm, over the past two years the time from online purchase to delivery has been reduced from 5.2 days to 4.3 days for most retailers, while Amazon is still fastest with an average of 3.2 days.[8]

Patrick Browne, UPS's head of global sustainability, makes the consequences clear. 'The time spent in transit has a direct relationship to the environmental impact. I don't think the average consumer understands the cost to the environment of having something tomorrow as opposed to two days from now. The more time you give me, the more efficient I can be.'[9] He was referring to trucks with only half a load or even a mere scattering of packages that must be delivered within a certain time. It is a fair bet that the Covid lockdowns and shuttered shops led to consumers insisting on still faster delivery times for their increased online purchasing. The impulse to acquire might arguably have become more insatiable of late, the swift arrival of the parcel being the entitlement of a customer who has already paid the money. It is not an attitude that favours slow steaming.

The Stuck Monkey Takes Stock

Squatting in the clearing with one paw in the jar, the Monkey dimly knows it is doomed. Sooner or later the hunters will return and that will be that. They may even be distracted by something else and not come back until the following day, but in the long run it will make no difference. Millions of years of evolution have brought the Monkey down from the trees and given it useful skills for getting termites out of holes by coating sticks with gum, but it has not given it the mental leap necessary to forgo immediate reward in favour of longer-term advantage. It is even further from realising that by letting go what after all is only a banana, it could go back to its troop and warn them never to put their paws in a pot, thereby probably guaranteeing its election as Boss Monkey.

At climate jamborees like COP26 the shrill schoolgirl saints and their disciples waving banners and screaming are now increasingly echoed by their young schoolteachers and parents. Their denunciations are full of righteous adolescent passion but are stupidly aimed at the soft targets of 'world leaders' who, as we jaded adults know, are the very last people able or wanting to change anything. Almost all of them are here-today-gone-tomorrow pragmatists, practically none of whom can guarantee to win their own re-election, let alone be able to change the world's entire economic system.

'We are in the beginning of a mass extinction and all you can talk about is the money and fairy-tales of internal economic growth,' Greta Thunberg squalled at the UN on 4 June 2019. The most scathing riposte to her sermon came from the veteran broadcaster Alan Jones of Sky News Australia. In his prime-time TV programme he was ostensibly addressing Australian teenagers, but what he said applied equally to their counterparts as well as to their parents everywhere in the developed world:

> You're the first generation who have required air conditioning in every classroom. You want TV in every room, and your classes are all computerised. You spend all day and night on electronic devices. More than ever you don't walk or ride bikes to school but you arrive in caravans with private cars that choke suburban roads and worsen rush-hour traffic. You're the biggest consumers of manufactured goods ever and update perfectly good expensive luxury items to stay trendy. Your entertainment comes from electronic devices. Furthermore, the people driving your protests are the same people who insist on artificially inflating the population growth through immigration which increases the need for energy, manufacturing and transport. The more people we have,

the more forest and bushland we clear, the more of the environment is destroyed.

How about this? Get your teachers to switch off the aircon. Walk or bike to school. Switch off your devices and read a book. Make a sandwich instead of buying manufactured fast food... No, none of this will happen because you're selfish, badly educated, virtue-signalling little turds inspired by the adults around you who crave a feeling of having a 'noble cause' while they indulge themselves in Western luxury and unprecedented quality of life. Wake up, grow up, and shut up!

At long last someone had understood exactly why the Monkey was stuck.

As this book has tried to make clear, there is no major technological breakthrough that has not had unforeseen consequences, all of which have been ruinous to the environment, often in unpredictable ways. However, there is always serious money to be made from them so they cannot be stopped. In many ways the unhindered proliferation of social media can now be seen as having been far more disaster than blessing, with unscrupulous or just plain ignorant people able to spread lies and mistrust worldwide. Their hatreds and falsehoods, popularly believed by the uneducated, not only eat away at the already threadbare fabric that still barely holds civilised society together but also steadily erode what we like to call 'reality' itself. In addition, well-rewarded influencers constantly promulgate fads and fashions that can have environmentally disastrous consequences. We ought not to be surprised by any of this. Our inability – or inertia or unwillingness – to prevent the next step from taking place before its likely outcomes have been properly thought through is merely business as humans have always done it. Not by bananas alone may Monkeys be caught. The same trick works even better with more glittery bait.

The next hi-tech step, Mark Zuckerberg's 'Metaverse', is the offspring of what has become known as platform capitalism, whereby those who own electronic platforms (Uber, Airbnb, etc.) need no expensive offices or physical infrastructure but amass fortunes merely by offering a service. However, this new step does require some very expensive equipment. At the moment what is billed as the world's fastest AI supercomputer is nearing completion. This, a gigantic array of electronic consoles arranged in sinister black cliffs lining an infinite recession of silent passages in an undisclosed location, is what will power the Metaverse. We dare not imagine how much power this humming behemoth will consume. Having renamed itself Meta, the social media platform Facebook will henceforth be accessible within the Metaverse to users wearing virtual reality (VR) and augmented reality (AR) headsets and 'smart' glasses. As the journalist Jonathan Chadwick wrote in studiedly neutral tones: 'This supercomputer could enable new AI systems that can power real-time voice translations to large groups of people in the Metaverse, each speaking a different language, so they can seamlessly collaborate on a research project or play an AR game together.'[1] No mention of the vast fortunes waiting to be made from product placement, advertising and retail. In the same article Louis Rosenberg, billed as 'one of the world's leading computer engineers', struck a note of warning over the Metaverse's likely effect on its participants. He said he thought that by integrating VR and AR and having people interact in the digital realm for a significant portion of their day, it could 'alter our sense of reality and distort how we interpret our direct daily experiences. Our surroundings will become filled with persons, places, objects, and activities that don't actually exist, and yet they will seem deeply authentic to us.'

Such a prediction seems plausible, given that so many people already opt to live much of their lives in cyberspace in various other worlds such as Second Life, Fortnite, Roblox and VRChat. In Zuckerberg's dystopian Orwellian future maybe people seen without their headsets and smart glasses will come to be outed and denounced, like American airline passengers who refused to wear face masks in the Omicron pandemic. Making the electronic illusion of reality still easier and more 'real' can only further blur any distinction between the worlds people inhabit. All the same, calling it 'augmented reality' does seem like a gross exaggeration. Your avatar's 'reality' will hardly be augmented if it can't smell or taste the food it's apparently eating, or feel the ground beneath its feet as it walks ('it' presumably meaning 'you'). It will be a severely diminished reality until – as promised – electronic wizardry catches up and is able to interfere with your brain to produce the illusion of taste, smell and touch. From then on, Wheee! The pornography! In the cyberspatial piece of real estate you own (and have paid for in hard cash back in dreary old 'reality') you install your ideal brothel, full of unlimited unlimitedness. And there you live out your days, lost in nightmares of ecstasy... Even so, no matter the banquets your avatar consumes in your chosen shadowland, you will still have to eat and shit in the real world just as you will still fall ill and die here as well.

Meanwhile, very much in real reality, the fabulously wealthy are planning their escape to live out their days in hutches made of beryllium (it's six times harder than steel, and lighter!) on Mars or the moon, able to go outside only when wearing protective suits and breathing bottled air for the rest of their lives. God knows what they will eat, and frankly who cares? It is all the perfect metaphor for the endgame in the story of Homo sapiens. Both Hunters and Monkey are choosing to desert planet Earth, the one by rocket and the other by illusion. It is a toss-up as to which destination will be the more unliveable but in both cases reality will become unignorable. In the case of the über-wealthy Hunters it will mean an ironic comedown almost to the status of hunter-gatherers on a world where there is nothing to hunt or gather, while living in cave-like conditions beside which quarters in a scientific station in Antarctica would seem like Club Med. What will they do in their Martian or Lunar exile apart from worrying endlessly about daily survival, solar radiation and meteorite strikes? Will they find themselves wistfully scanning a cindery horizon for a glimpse of anything other than cosmic geology?

But no – it won't be like that at all. These start-up trillionaires will be living in the alternative reality they themselves invented and enabled. Their physical selves may be unable to go for a walk without wearing diapers in a spacesuit and carrying bottles of oxygen, but their avatar selves will still be busily carrying on with partying, making deals, dating, designing cities, buying stuff and all the other things the digital mirage makes available. They may have to crap in their spacesuits as they trudge across moondust or Martian desert but their minds will be free to roam their ersatz worlds, making the same old blunders and ill-advised decisions as before, leaving behind botched relationships and all the rest of what it so messily means to be human. How otherwise? And the worlds on which they are encamped that did for once correctly earn the description 'pristine' will inevitably, very slowly, become polluted by their presence.

And the Monkey? The jungle was long ago felled for a beef ranch, and the former clearing in which the Monkey once sat with its paw in a jar was for a while trodden by cattle hooves. But the venture soon failed, the Earth baked to grassless sterility by climate change. Well before then the Hunters had deserted their parched village in search of water.

Appendix 1: Petroleum-based Products in Everyday Life

(see https://whgbetc.com/petro-products.pdf)

Other than the specific chemicals derived from the petroleum industry mentioned in the text (ammonia, methanol, etc.), the following arbitrary selection lists a tiny fraction of the innumerable products that are ultimately made, or derive, from oil. Even if one leaves aside petroleum-based energy (petrol, jet fuel, heating oil, etc.) the modern world consists to an extraordinary degree of things that owe their existence to Big Oil. Virtually all plastics are made from petroleum and they and their derivatives appear almost everywhere. Among thousands of oil-related products are:

Paraffin wax, fertilisers, pesticides, herbicides, detergents, long-playing records, photographic film, furniture, packaging materials, surfboards, paints, artificial fibres used in clothing, upholstery, curtains, carpets, nightdresses, solvents, cassettes, motorbike helmets, CD players, vitamin capsules, putty, percolators, skis, tool racks, mops, umbrellas, roofing, denture adhesive, loudspeakers, tennis racquets, nylon rope, water pipes, shampoo, guitar strings, antifreeze, combs, vaporisers, heart valves, nail varnish, anaesthetics, dentures, cold cream, fan belts, refrigerators, toys...

diesel, asphalt, floor wax, car bodies, caulking, tap washers, food preservatives, cortisone, dolls, tents, shoes, glues, inks, cameras, dice, handbags, pillows, antihistamines, life jackets, TV cabinets, computers, car battery cases, toilet seats, linoleum, plastic wood, rubber cement, candles, hand lotion, suitcases, toothbrushes, CDs, balloons, crayons, artificial grass, cinema film, lacquer, golf balls, washing-up liquid, unbreakable dishes, sound insulation, electrical tape, bicycle and car tyres, contact lenses, disposable nappies, marker pens, foam rubber, boat hulls, antifreeze...

biros, Vaseline, antiseptics, basketballs, deodorants, tights, rubbing alcohol, shag rugs, epoxy, insect repellents, fishing rods, ice cube trays, gum boots, bin liners, aspirin, awnings, paintbrushes, paint rollers, sunglasses, parachutes, artificial limbs, shaving cream, toothpaste, perfumes, shoe polish, clotheslines, soap, oil filters, hair dye, lipstick, glycerine, skateboards, surfboards, shower curtains, aircraft seats, cameras, bandages, hair curlers, picnic mugs, ammonia, hearing aids, paddling pools, mobile phones, electric plugs, wood varnish, transfusion kits, duct tape, piano keys...

And so on, ad infinitum.

Appendix 2: Biofuels

The chapter on cars reveals some of the reasons why neither battery nor hydrogen power is likely to solve our carbon-emitting transport problem as we wait for technology and logistics to catch up. This present chapter will show why biofuels, while seeming to present an interim Green alternative for powering internal combustion engines, are both insufficient and themselves destructive of the global environment.

When buses, cars or trucks are powered by a biofuel – typically biodiesel – most people quite reasonably think that it sounds more Green than a fossil fuel pumped out of the ground by Big Bad Oil. A moment's reflection will reveal a deep irony here since of course petroleum, too, is a biofuel in the sense that like coal, it was formed by biomass: prehistoric plants, forests and animal corpses converted by millions of years' geological heat and pressure into oil. (Much the same could be said of plastics since they, too, start as crude oil, reach an apotheosis as Coca-Cola bottles and may, with ingenious chemistry and a good deal of energy, be reduced back to oil, albeit with significant loss.)

One difference between a biofuel and petroleum is that biofuels are a renewable energy resource and petroleum is not. More importantly, a biofuel is slightly cleaner. America's Environmental Protection Agency (EPA), has shown that biodiesel emits 11 per cent less carbon monoxide and 10 per cent less particulate matter than standard diesel, while the Department of Energy and Agriculture (DEA) found that biodiesel reduces net CO_2 emissions by an encouraging 78 per cent. An additional advantage is that biofuels can be made from organic waste such as used cooking oil and even, presumably, fatbergs.

ETHANOL

As it happens, a percentage of practically all the petrol sold at pumps throughout Europe, the UK and North America already contains a biofuel. One of the problems with petrol is that it needs additives in order to function smoothly in highly tuned and modern engines. This is why lead used to be added until it was banned as a dangerous pollutant in the 1980s. After that, the most common fuel additive was something called methyl tert-butyl ether (MTBE) until it was identified as a potential carcinogen. What was Big Oil to do? Its chemists decided that ethanol – and to a lesser extent methanol – fitted the bill admirably as an additive, and even more so because if it was produced from a crop like maize or sugarcane it could be classed as a biofuel and hence as Green. Ethanol is ethyl alcohol (C_2H_6O) – the eminently drinkable alcohol that gives gin, beer and any other alcoholic drink its kick. Methanol or methyl alcohol (CH_3OH) is chemically similar but poisonous. It kills hundreds – perhaps even thousands – of people each year worldwide, usually as the result of botched attempts to distil homebrewed liquor, resulting in the wrong kind of alcohol. As the name suggests, it is the basis of methylated spirit. Both ethanol and methanol can be made from biomass and ethanol has long been a major source of power for cars and lorries in Brazil. It can be distilled from all sorts of grain and waste material but in that country is most commonly sourced from sugarcane.

ETHANOL/PETROL

An ethanol/petrol mix has proved very successful. The United States is the world's largest fuel ethanol producer (Brazil is the second largest) and nowadays most of the petrol sold to US, UK and EU car drivers contains 5 or 10 per cent ethanol, mixes usually known at the pumps as E5 and E10. This petrol/alcohol blend burns cleaner than would pure petrol, with less carbon dioxide in the exhaust. There are drawbacks, however. Ethanol itself produces 34 per cent less energy than petrol and is less volatile, so cold engine starting can sometimes be slightly more difficult. On combustion the ethanol/petrol mix also produces water, which can lead to more engine wear. Even at the low concentration of 10 per cent the fuel may also damage older car engines. A higher ethanol concentration as E15 can be used in cars built after 2001. Flexible fuel vehicles (FFVs) that can handle petrol with 85 per cent ethanol (sold as E85) are becoming more common. This mixture is definitely not suitable for ordinary cars because the high ethanol content causes the rubber and plastic parts of their fuel systems to perish. However, because they are specifically built for E85, FFVs can have better performance in terms of acceleration and horsepower while producing fewer emissions.

METHANOL

Methanol has many of the same properties as ethanol as a fuel additive and like ethanol can cut CO_2 emissions in the exhaust by up to 95 per cent and nitrogen oxides (NOx) by 80 per cent. It is cheaper to produce sustainably than ethanol but has a lower energy density. Another drawback is that not only can it be fatal if mistakenly drunk but its vapour is poisonous and readily absorbed through skin and lungs. As a chemical building block it has a host of uses for making industrial and home products. As the main source of formaldehyde (formalin) methanol goes to make the ubiquitous fibreboard known as MDF as well as all sorts of medical equipment including gloves, masks, gowns, disinfectants and pharmaceuticals. Like ethanol, methanol can be obtained by the gasification of biomass or directly from the methane in natural gas. It is also easily synthesised from a mix of gases: hydrogen, carbon dioxide and carbon monoxide, all of which can be sourced from a variety of biomass feedstocks. However, because of its poisonous vapour most countries set its limit in domestic petrol at 3 per cent.

BIODIESEL

All this means that the petrol available to most motorists is not itself a biofuel since it is still 95 per cent a fossil fuel and only one-twentieth part Green. Much more truly Green is 100 per cent biodiesel which has no fossil fuel content and is identifiable at the pump as B100 although the vehicle's engine needs to have been built from scratch to run on it. As with ethanol, biodiesel can be blended with varying proportions of Big Oil's petrodiesel and most diesel engines will run on a B20 mix (i.e., 20 per cent biodiesel) without modification. True B100 biodiesel can be made from biomass, oils from oil palms and soybeans and also from waste oils such as cooking fat. All these sources require some fancy but well-understood industrial chemistry to produce biodiesel.

BIOMASS AND BIOENERGY CROPS

On the face of it biofuel's Green credentials sound pretty good. Unfortunately they are not as good as they sound unless entire national agricultural policies are changed to grow and collect the required types and quantities of feedstocks, together with the necessary infrastructure and industrial plant. They are promoted internationally by Big Oil itself, being the one industry that has the best knowledge of fuels and their chemistry and already capable of synthesising just about anything from anything else – at a price. Biofuel is currently promoted internationally by all sorts of laws and incentives. One of the most trumpeted sources is biomass, meaning almost any renewable plant waste, from grass cuttings to large trees. Biomass is usually billed as carbon-neutral, meaning that the carbon the plants absorbed from the atmosphere via photosynthesis while they were growing is simply given off again when they are burned, so overall no carbon is gained and none lost. Recently a good deal of research has queried this optimistic assertion and much depends on the type of biomass. Burning large trees has been shown to generate slightly more carbon than they sequestered in the first place. In any case there is a limit to the amount of usable biomass that is globally available, and it can never be anything like enough to supply our needs for 100 per cent biofuels. As it is, the quest for biofuels is already doing immense damage to the cause of biodiversity.

By far the majority of biofuels are not made from organic waste but from purpose-sown 'bioenergy' crops such as soya, oil palm, maize and sugarcane. Thanks to satellite surveying we can see what is happening on a day-to-day basis. The old days when governments could lie-and-deny about what was going on inside their territory have gone for good. Huge areas of forest have been – and are being – visibly felled in countries like Indonesia, Malaysia and Brazil in order to plant countless hectares of oil palms. Elsewhere, existing farmland has been converted to oil palm plantation, forcing replacement farmland to be created from virgin forest. The net damage is the same. For this reason palm oil has become a target of demonstrations in recent years and many people boycott products that contain it. This is often extremely hard to do since the palm oil boom has rendered it the world's most widely used vegetable oil. The World Wildlife Fund claims palm oil has now found its way into nearly half of all consumer products, including cosmetics, soap, detergents, and processed foods of all kinds from ice cream to pet food and, of course, those 'health-packed' protein bars.

In the last six years the demand for vegetable oils has greatly increased. In Europe over half of all imported palm oil is converted into biofuel for vehicles. With the aviation industry also keen to 'go Green' this trend can only grow, further shrinking the world's fast-dwindling forests. Thanks to those satellites we know that in 2020 alone a further 4.2 million hectares of virgin rainforest were cleared for agriculture (principally palm oil and soya) and cattle. Brazil alone lost 1.7 million hectares – an astonishing 6,563 square miles, almost all of it for beef production, a large proportion of which will find its way into pet food. Such losses of tree cover have a massive effect on the planet's climate. The deforestations in 2020 were alone equivalent to adding some 2.6 billion tons of CO_2 to the atmosphere: about the same amount as India's annual output as the world's fourth-largest CO_2 emitter. Quite by chance, that almost exactly negates all the gains made by global tree-planting campaigns up until now.

As with all biofuels, ethanol carries its own environmental penalties. Sugarcane production contaminates rivers with silt and fertilisers that between them often cause offshore damage (parts of the Great Barrier Reef are typical casualties). Merely to produce sugar for human consumption requires large volumes of water. A single pound of refined sugar takes 213 gallons of water to produce, or around three gallons per teaspoonful. Thanks to sugarcane farming Brazil's Atlantic Forest now covers barely 7 per cent of its original area and that country's ethanol production accounts for 39 per cent of the world's supply. Sugarcane cultivation alone now covers 26.3 million hectares worldwide. In 2015 one academic paper found it would take 'about twenty years for the lower emissions from [vehicular] sugarcane ethanol to "pay back" the added emissions from deforestation.'[1] The huge and ever-growing demand for sugar in processed food of every kind is a separate issue with its own tally of Earth's resources.

In the United States, maize (or corn, as it is known there) was proposed as a widely grown potential source of ethanol. In 2014 the Intergovernmental Panel on Climate Change encouragingly confirmed that 'Biofuels have direct, fuel-cycle greenhouse gas emissions that are typically 30 to 90 per cent lower than those for petrol or diesel.' But that estimate was reached without taking into account the indirect effects on ecosystems and biodiversity of destroying forests and changing land usage into monocultures. The last word should probably go to the International Institute for Sustainable Development, which has estimated that the climate benefits from ethanol are zero. It adds that in the US it would be a hundred times more effective, as well as far cheaper, simply to lower vehicle emissions and increase corporate average fuel economy (CAFE) standards on all cars and light trucks to a minimum of forty miles per gallon, as in Japan. It is hard to see this happening in the land where 'personal freedom' is so idiosyncratically defined.

For British readers it should be added that just because the UK is banning the sale of new petrol- or diesel-powered vehicles after 2030 they should not imagine that millions of existing polluting vehicles will disappear from the roads overnight. They won't. But even as they do slowly vanish, the chances are that many of them will not be scrapped but will find their way overseas to hotter countries where vehicular emissions are not so strictly policed.

Endnotes

How the Monkey Got Stuck

1. Clarkson, N., 'Amazon founder Jeff Bezos's £48m Gulf Stream has led a 400-strong parade of private jets into COP25', *Daily Mail Online* (1 Nov. 2021).
2. Crawford, N. C., Watson Institute for Public Affairs, Brown University, 2019 figure. https://watson.brown.edu/costsofwar/papers/ClimateChangeandCostofWar; accessed 18 Oct. 2022.
3. Wallace-Wells, D., 'Ten Million a Year', *London Review of Books*, 43:23.
4. Subin, A. D., *Accidental Gods* (Cambridge: Granta, 2021), pp. 133–4.

Eat Your Pets!

1. Horowitz, K., 'All That Meat in Pet Food Has a Big Environmental Impact', *Mental Floss*, 4 Aug. 2017, www.mentalfloss.com/article/503376/all-meat-pet-food-has-big-environmental-impact; accessed 3 Dec. 2021.
2. https://thedogvisitor.com/; accessed 5 Dec. 2021.
3. BBC News, 1 Jan. 2022.
4. King, J., 'No traces of TB found in Geronimo the alpaca's dead body', *Guardian* (10 Dec. 2021).
5. Horton, I. and Abiona, J., 'Why that lockdown hot tub is money down the drain: Thousands of Britons now regret splashing out on luxury items such as pizza ovens and gym equipment during

Covid shut-ins, research shows', *Daily Mail* (8 Nov. 2021).

6. *Guardian* (3 Jan. 2022).
7. BBC News, 23 Oct. 2021.
8. www.mentalfloss.com/article/503376/all-meat-pet-food-has-big-environmental-impact
9. https://www.pdsa.org.uk/pet-help-and-advice/looking-after-your-pet/puppies-dogs/the-cost-of-owning-a-dog; accessed 21 Oct. 2021.
10. www.gminsights.com/industry-analysis/pet-care-market; accessed 19 Oct. 2022.
11. www.statista.com; accessed 8 Dec. 2021.
12. Pirie, T. J., Thomas, R. L. and Fellowes, M. D. E., 'Pet cats (Felis catus) from urban boundaries use different habitats, have larger home ranges and kill more prey than cats from the suburbs', *Landscape and Urban Planning*, Volume 220, Article 104338.
13. Gorvett, Z., 'The hidden reason processed pet foods are so addictive', *BBC Future, Beyond Natural*, May 2021, https://www.bbc.com/future/article/20210519-the-hidden-reason-processed-pet-foods-are-so-addictive
14. Ly, C., 'Domestic cats are driving parasitic infections in wild animals', *New Scientist* (30 Oct. 2021).
15. Ly, C., 'Dog waste may harm nature reserve biodiversity by fertilising the soil', *New Scientist* (12 Feb. 2022).
16. freshwaterblog.net; accessed 17 Nov. 2020.
17. AFP, 'Raw dog food "may be fuelling spread of antibiotic-resistant bacteria"', *Guardian* (10 Jul. 2021).
18. futuremarketinsights.com; accessed 10 Mar. 2021.

Gardening

1. *Private Eye*, issue 1561 (26 Nov.–9 Dec. 2021).
2. Gillam, C., 'Only two out of 11 herbicide studies deemed reliable', *Guardian* (26 Nov. 2021).
3. Carrington, D., 'Peat sales to gardeners in England and Wales to be banned by 2024', *Guardian* (18 Dec. 2021).

Sport

1. Richards, G., 'Climate emergency accelerates F1's efforts to clean up its image', *Guardian* (26 Nov. 2021).
2. Stanton, J., Lockwood, D. and Gornall, K., 'Premier League. Should clubs stop flying to domestic matches?' *BBC Sport* (12 Nov. 2021).
3. Benson, A., 'Formula 1 launches a plan to become carbon neutral by 2030', *BBC Sport* (12 Nov. 2019).
4. Berners-Lee, M., 'What's the carbon footprint of ... the World Cup?' *Guardian* (10 Jun. 2010).
5. 'Winter Olympics: Artificial snow could cause environmental damage – report', *BBC News* (26 Jan. 2022).

Cars & Aircraft: Hybrid, Electric and Hydrogen-powered

1. Read, B., 'Testing sustainable technology', *Aerospace* (Dec. 2021).
2. Dron, A., 'Fuelling aviation sustainability', *Aerospace* (Jan. 2022).
3. Carrington, D., 'Car tyres produce vastly more particle pollution than exhausts, tests show', *Guardian* (3 Jun. 2022).
4. Schätzl, S., *Kronen Zeitung* (22 Dec. 2021).
5. Vaughan, A., 'A hydrogen fuel revolution is coming', *New Scientist* (3 Feb. 2021).
6. Uekckerdt, F., 'Potential and risks of hydrogen-based e-fuels in climate change mitigation', *Nature* (6 May 2021).
7. Harper, G., 'Recycling lithium-ion batteries from electric vehicles', *Nature* (6 Nov. 2019).
8. Widburg, A., 'Hunter Biden, China, electric batteries, and child slavery', *American Thinker* (21 Nov. 2021).
9. Coward, R., 'Streets ahead? What I've learned from my year with an electric car', *Guardian* (8 Jan. 2022).
10. BBC Cumbria News, 12 Dec. 2021.
11. Hadaway, S., 'RAF Fauld and the RAF munitions supply network', *RAF Historical Society Journal* 78 (2022), p. 28.

The Fashion Industry

1. UN Climate Change News, 6 Sept. 2018.
2. *Geographical* (Mar. 2021).
3. Australian Government figures: Department of Agriculture, Water and the Environment, 2021.
4. Syren, F., 'Cotton – one of the most pesticide-contaminated crops in the world', 1 Dec. 2017, zerowastefamily.com (20 Oct. 2022).
5. Author's lockdown interview with fieldwork student, 23 Nov. 2021.
6. www.oceancleanwash.org; accessed 20 Oc. 2022.
7. Pitcher, L., 'New Study links major fashion brands to Amazon deforestation', *Guardian* (30 Nov. 2021).
8. digitaltrends.com; accessed 20 Oct. 2022.
9. Hale, B., 'Shameful cost of your fast fashion', *Daily Mail* (28 Jan. 2022).
10. BBC News, 8 Oct. 2021.

Military Carbon

1. Ambrose, T., 'World's militaries avoiding scrutiny over emissions, scientists say', *Guardian* (11 Nov. 2021).
2. Sabbagh, D., '"Colossal waste". Nobel laureates call for 2% cut to military spending worldwide', *Guardian* (14 Dec. 2021).
3. Distel, Marshall essay on fossilfuel.com; accessed 15 Mar. 2020.
4. Ambrose, T., 'World's militaries avoiding scrutiny over emissions, scientists say', *Guardian* (11 Nov. 2021).
5. Neimark, B., Belcher, O., Bigger, P., 'US Military Pollution', *The Ecologist* (27 Jun. 2019).
6. Watson Institute for Public Affairs, Brown University, https://watson.brown.edu/costsofwar/; accessed 5 Jan. 2022.
7. Distel, Marshall essay on fossilfuel.com; accessed 15 Mar. 2020.
8. Karbuz, S., 'US Military Energy Consumption: Facts and Figures', https://www.resilience.org/stories/2007-05-21/us-military-energy-consumption-facts-and-figures/; accessed 7 Jan. 2022.

9. International Renewable Energy Agency: 'Biofuels for Aviation', https://www.irena.org/publications/2017/Feb/Biofuels-for-aviation-Technology-brief; accessed 7 Jan. 2022.

10. Davies, C., worldbeyondwar.org, (2020); accessed 7 Jan. 2022.

11. Pirani, S., *Burning Up: A Global History of Fossil Fuel Consumption* (London: Pluto Press, 2018).

12. Bonnett, A., *The Age of Islands* (New York: Atlantic 2020), p. 43.

13. Dahr, J., 'Iraq's wars, a legacy of cancer', https://www.aljazeera.com/features/2013/3/15/iraq-wars-legacy-of-cancer

14. Hambling, D., 'US arms maker ends production of controversial depleted uranium rounds', *New Scientist* (19 Feb. 2022).

Happy Holidays: Eco-lodges and Cruises

1. https://www.maharajaecodivelodge.com; accessed 10 Dec. 2021.

2. Randall, I., 'Scientists studying microplastics in Antarctica', *Daily Mail* (13 Dec. 2021).

Mobile Phones & Computers

1. Berners-Lee, M., 'What's the carbon footprint of ... using a mobile phone?' *Guardian* (9 Jun. 2010).

2. Adams, W. M., www.datacenterdynamics.com/en/opinions/power-consumption-data-centers-global-problem/; accessed 3 Feb. 2022.

3. *Private Eye* (12 Nov. 2021).

4. Globalwaste.org; accessed 10 Mar. 2022.

Wellness & Beauty

1. Kale, S., 'Get shredded in six weeks!' *Guardian* (27 Jun. 2018).

2. Fisher, J., BBC.com, 15 Feb. 2022, quoting a York University study: 'Pharmaceutical Pollution of the World's Rivers', PNAS (2022).

3. Fuller, G., 'How much indoor air pollution do we produce when we take a shower?' *Guardian* (17 Dec. 2021).
4. *Miami Herald* (3 May 2021).
5. *Washington Post* (26 Dec. 2021).

Personal Freedom vs the Planet: Cryptocurrencies

1. Shurk, J. B., 'The Four Horsemen of the Left's Artificial Apocalypse', *American Thinker* (7 Jan. 2022).
2. 'Hillary Clinton: "Cryptocurrencies have potential of undermining the dollar"', *Breitbart* (19 Nov. 2021).
3. Davies, R., 'Trading is gambling – no doubt about it', *Guardian* (15 Jan. 2022).
4. Lu, D., 'Bitcoin mining emissions in China will hit 130 million tonnes by 2024', *New Scientist* (17 Apr. 2021).
5. Milman, O., 'Bitcoin miners revived a dying coal plant', *Guardian* (18 Feb. 2022).
6. '"Tesla will no longer accept Bitcoin over climate concerns," says Musk', *BBC News Online* (13 May 2021), https://www.bbc.co.uk/news/business-57096305
7. Ibid.
8. Walker, B., 'At Satoshi's Tea Garden', *London Review of Books* (43:9), p.37.
9. de Vries, A. and Stoll, C., 'Bitcoin's growing e-waste problem', Resources, Conservation and Recycling, 153 (2021), https://doi.org/10.1016/j.resconrec.2021.105901
10. Wintour, P., 'No. 10 pressured me to drop anti-money laundering measures, says ex-minister', *Guardian* (15 Feb. 2022).

Shipping & Shopping

1. Griffiths, G. of S&P Global Platts, quoted Baraniuk, C., 'Why even giant ships can't solve the shipping crisis', *BBC News Online* (14 Sept. 2021).
2. Lanchester, J., 'Gargantuanisation', *London Review of Books* (43:8), p. 2.

3. Stratiotis, E., 'Fuel Costs in Ocean Shipping', https://www.morethanshipping.com/fuel-costs-ocean-shipping/; accessed 10 Mar. 2022.
4. Sparkes, M., 'Drones are "sniffing" ship exhaust for illegal fuel in European waters', *New Scientist* (9 Oct. 2021).
5. Woodcock, A., 'Brexit shift in trade away from EU "could double UK carbon emissions from shipping"', *Independent* (29 Oct. 2021).
6. Jones, N., 'Swabbing the decks', *New Scientist* (27 Nov. 2021).
7. Cleary, B., 'American teacher shares what shocked her about shopping in Kmart in Australia', *Daily Mail Online Femail* (26 Oct. 2021).
8. Choudhary, A., 'The Environmental Cost of Shipping: A Brief Explanation', 4 Aug. 2021, letclotheslivelong.org
9. Ibid.

The Stuck Monkey Takes Stock

1. Chadwick, J., 'Meta is building the world's fastest AI SUPERCOMPUTER', *Daily Mail* (24 Jan. 2022).

Appendix 2: Biofuels

1. Sant'Anna, Marcelo, 'How green is sugarcane ethanol?', https://direct.mit.edu/rest/article-abstract/doi/10.1162/rest_a_01136/108835/How-Green-Is-Sugarcane-Ethanol?redirectedFrom=fulltext; accessed 22 Mar. 2022.

Index

Entries, other than names/proper nouns, containing initial capitals indicate headings and sub-headings in the text.